T0276065

SpringerBriefs in Applied Sciences and Technology

More information about this series at http://www.springer.com/series/8884

Victor V. Kozlov · Genrich R. Grek
Yury A. Litvinenko

Visualization
of Conventional and
Combusting Subsonic
Jet Instabilities

 Springer

Victor V. Kozlov
Khristianovich Institute of Theoretical
  and Applied Mechanics
Siberian Branch of Russian Academy
  of Sciences
Novosibirsk
Russia

Yury A. Litvinenko
Khristianovich Institute of Theoretical
  and Applied Mechanics
Siberian Branch of Russian Academy
  of Sciences
Novosibirsk
Russia

Genrich R. Grek
Khristianovich Institute of Theoretical
  and Applied Mechanics
Siberian Branch of Russian Academy
  of Sciences
Novosibirsk
Russia

Additional material to this book can be downloaded from http://extras.springer.com.

ISSN 2191-530X   ISSN 2191-5318   (electronic)
SpringerBriefs in Applied Sciences and Technology
ISBN 978-3-319-26957-3   ISBN 978-3-319-26958-0   (eBook)
DOI 10.1007/978-3-319-26958-0

Library of Congress Control Number: 2015956132

Printed on acid-free paper

This Springer imprint is published by SpringerNature
The registered company is Springer International Publishing AG Switzerland

Preface

Stability of jets and combustion of jet fuels are the focus of long-term studies due to both fundamental and engineering aspects of these phenomena. A wealth of research data on this topic is available today and reported in original journal papers, reviewing articles, and monographs. In the present e-book, we consider, most of all, the influence of initial conditions at a nozzle exit and acoustic effects on round and plane, macro- and microjets at low subsonic velocities, according to our experience in experimental work on this subject. We expect that the material presented here will be of interest to high school teachers, university students, and researchers engaged in jets-related problems.

Organization of the Book

In Chap. 1, we present a review of the main works devoted to the studies of conventional and combusting subsonic jet instabilities.

In Chap. 2, we begin with the dynamics of a round jet emphasizing the contribution of longitudinal disturbances to the perturbed flow pattern. Important issues are generation, spatial development, and interaction of three-dimensional streaky structures with ring vortices.

In Chap. 3, we proceed with plane jets, focusing again on the formation and development of longitudinal structures of laminar flow disturbances and their interaction with the Kelvin–Helmholtz vortices. We than present some visualization results on a free jet, followed by a detailed consideration of a perturbed wall jet.

Obviously, evolution of the jet depends on the initial conditions at the nozzle exit. In particular, varying nozzle geometry and, hence, the velocity distribution near its exit, it is possible to modify instability of a jet and its dynamics. Such an approach to control of round jets is the main subject of Chap. 4.

In Chap. 5, laminar and turbulent round jets are under further consideration. Experimental data testify to the similarities of the generation and development of coherent structures in both cases. Also, almost one and the same response of jets to external acoustic oscillations is demonstrated.

The influence of initial conditions at a nozzle exit and acoustic perturbations on a plane jet structure, its evolution and stability, is considered in Chap. 6. We demonstrate basic features of laminar and turbulent plane jets at one and the same Reynolds number. It is found that jets are subjected to sinusoidal oscillations suppressing the varicose mode of instability. The focus here is the interaction of the longitudinal structures of flow perturbations generated at one side of the nozzle, with the large-scale two-dimensional vortices of the laminar plane jet resulting in origination of Λ- or Ω-shaped vortex structures.

In Chap. 7, we present experimental and numerical data, compiled by studying characteristics of round jets with a top hat and a parabolic mean velocity, which exhibit profiles at the nozzle exit in a cross-flow. In the case of parabolic velocity distribution, flow instability results in the jet deformation appearing as tangential bursts of fluid so that a pair of counter rotating stationary vortices is generated and the jet core is diminished. As is found, the jet/cross-flow interfaces might have undergone a "stretching and thinning" process caused by the cross-flow.

In Chap. 8, we consider the effect of a transverse acoustic field on round and plane, macro- and microjets at small Reynolds numbers.

Finally, Chap. 9 focuses on the jet flames at low Reynolds numbers. When forcing propane jets by transverse acoustic oscillations, new features of combustion were observed and explained.

Supplementary Material

Most of the above listed chapters are supplemented by multimedia files providing visual illustrations of the phenomena discussed in the body of the present book.

Acknowledgments

A large part of the original experimental results shown in what follows was obtained in collaboration with our colleagues M.M. Katasonov, O.P. Korobeinichev, G.V. Kozlov, M.V. Litvinenko, N.I. Rudenko, A.G. Shmakov, and A.M. Sorokin that is highly appreciated. Also we are grateful to A.V. Dovgal, A.V. Boiko, V.G. Chernoray, L. Löfdahl, D.M. Markovich, and V.I. Zapryagaev for helpful suggestions.

Victor V. Kozlov   *Genrich R. Grek*   *Yury A. Litvinenko*

Novosibirsk
November 2015

Contents

Nomenclature

Only Main Symbols

Re	Reynolds number
U	Mean velocity
d	Nozzle diameter
x, y, z	Cartesian coordinates
ADC	Analog-to-digital converter
PC	Personal computer
RMS	Root-mean-square velocity
u'	Fluctuation velocity
Λ	Λ-like vortex structure
Ω	Ω-like vortex structure
2D	Two-dimensional
3D	Three-dimensional
f	Frequency
Hz	Hertz
kHz	Kilohertz
dB	Sound decibel intensity
l	Length
h	Height
v	Kinematic fluid viscosity
λ	Wavelength
Sh	Strouhal number
r	Current nozzle radius
R	Nozzle radius

K	Ratio parameter
δ	Boundary layer scale
φ	Equivalence ratio

Subscripts

0	Refers to initial value
∞	Refers to infinity
d	Refers to nozzle diameter
h	Refers to nozzle width
i	Refers to i-point velocity
jet	Refers to jet velocity
max	Refers to maximal velocity
mean	Refers to mean velocity

Chapter 1
Introduction

The focus of the present book is instability phenomena in subsonic jets which are of much interest both from fundamental and practical viewpoints. Mean velocity profiles of the jet shear layers are inflectional ones resulting in the Kelvin–Helmholtz vortices. The initial stage of their evolution is well predicted by the linear stability theory. The following nonlinear events at the laminar jet breakdown are characterized by saturation of the vortices and their amalgamation. Further development of the nonlinear perturbations is accompanied by emergence of streamwise elongated disturbances which are often due to a secondary three-dimensional instability of the Kelvin–Helmholtz vortices. Longitudinal disturbances evolving in jets can also be generated at non-uniformities of the nozzle surface. Most of all, they represent stationary or quasi-stationary three-dimensional deformations of the streamwise mean velocity and, under appropriate conditions, can experience non-modal growth as a transient effect of hydrodynamic instability. Such perturbations, appearing at flow visualization as streaky structures (streaks), are subjected to intensive interaction with oscillations of the laminar flow, for example, with instability waves that normally accelerates the transition to turbulence.

Obviously, the above features are important for the mixing process affecting the combustion efficiency, heat transfer, environmental pollution, and the jet noise. The longitudinal structures and streamwise vortices generated in a jet by different means in addition to the Kelvin–Helmholtz vortices may have a significant influence upon the mixing rate (Hu et al. 2001). Particularly, a lobed nozzle used for this purpose (Zaman 1997) appears as a promising tool of the mixing enhancement and is widely applied to control of the exhausts of turbofans and ejectors. With such lobed nozzles/mixers the jet noise and fuel consumption are reduced in engines of aircraft by Boeing and Airbus (Kuchar and Chamberlin 1980; Presz et al. 1994). Then, the infrared radiation of a battle plane is minimized ensuring its survival while accelerating mixing of the high-temperature, high-speed jet of the engine with the surrounding cold air (Power et al. 1994; Hu et al. 1999). The lobed nozzles are also used in Tiger (Germany, France), and Comanche (United States) helicopters, and F-117 stealth aircraft. Such nozzles were applied for the enhancement of mixing of

© The Author(s) 2016
V.V. Kozlov et al., *Visualization of Conventional and Combusting Subsonic Jet Instabilities*, SpringerBriefs in Applied Sciences and Technology, DOI 10.1007/978-3-319-26958-0_1

fuel and air in combustion chambers promoting the efficient combustion and reduction of contaminants, as well (Smith et al. 1997). For the first time, the dynamic mechanism of the mixing process behind a lobed nozzle/mixer was studied by Paterson (1984). It was shown that such a nozzle generates longitudinal vortices of the length of about one nozzle radius. A more detailed flow pattern in the jet of a lobed mixer was presented by McCormic and Bennett (1994). As is found, the interaction of the Kelvin–Helmholtz and the longitudinal vortices leads to a high degree of mixing. The longitudinal vortices transform the Kelvin–Helmholtz ones into "wrinkled" structures and enhance the mixing process. It turns out that the large-scale longitudinal vortices artificially generated at a lobed nozzle together with the Kelvin–Helmholtz vortices play an important role in the process of the jet core mixing with the surrounding fluid.

While in the above references the longitudinal vortices were generated by a special geometry of the nozzle, those originating from secondary instabilities of the jet as well as their contribution to the mixing process were investigated by Demare and Baillot (2001). Similar perturbations have been examined in plane shear layers (Bernal and Roshko 1986; Lasheras et al. 1986) and observed in round water jets (Liepman and Gharib 1992). The results of Lasheras et al. (1986) indicated that the origination of longitudinal vortex structures is a response of the shear layer to three-dimensional disturbances in upstream flow sections. According to Monkewitz et al. (1989) and Monkewitz and Pfizenmaier (1991), the formation of side jets in a hot jet is also a result of the development of longitudinal vortices. The presence of secondary vortices has been revealed through direct numerical simulation of round jets (Abid 1993; Brancher et al. 1994) and shear layers (Metcalfe et al. 1987).

Noteworthy is that the vortex structures in jets are strongly receptive to external noise. Therefore, using controlled acoustic disturbances a considerable variation of the jet characteristics can be achieved (Crow and Champagne 1971), in particular, for reduction of the turbulence level (Zaman and Hussain 1981). Acoustic perturbations can be applied to modification of the combustion process and reduction of the smoke black formation and nitric oxide emission in the jet burners (Chao et al. 1996). The experiments of Demare and Baillot (2001) showed that acoustic forcing of the jet enhances mixing and accelerates the stabilization of combustion. Detailed studies on the control of jets using acoustic excitation are reported by Guinevskiy et al. (2001).

In the work by Bagheri et al. (2009) it is marked, that the generic flow configuration of a jet in cross flow is ubiquitous in a great variety of industrial applications, ranging from the control of boundary-layer separation to pollutant dispersal from chimneys, from film cooling of turbine blades to the injection of fuel into combustion chambers and furnaces. The flow structures, mixing properties and general dynamics of jets in cross flow have therefore been the subject of numerous experimental and computational studies. In general four main coherent structures (see e.g. Fric and Roshko 1994; Kelso et al. 1996; Muppidi and Mahesh 2007 and the references therein) characterize the jet in cross flow: (i) the counter-rotating vortex pair, which originates in the near field of the jet and essentially follows the jet trajectory and dominates the flow field far downstream; (ii) the shear-layer vortices which are located at the upstream side of the jet and take the form of ring-like or loop-like filaments; (iii) horseshoe vortices forming in the flat-plate boundary layer upstream of the jet exit and corresponding wall vortices downstream of the exit close to the wall; and (iv) "wake vortices/upright vortices" which are vertically oriented shedding vortices in the wake of the jet. The accurate description of these relevant features is a prerequisite for a sound understanding of the perturbation dynamics of jets in cross flow and a first step in an attempt to manipulate it.

Recent advances in computational methods have enabled global stability analyses of flows with nearly arbitrary complexity and have furnished the possibility to assess fully two- and three-dimensional base flows as to their stability and response behaviour to general three-dimensional perturbations. Specifically, the combination of new efficient methods for computing steady-state solutions has provided the necessary tools for an encompassing study of the disturbance behaviour in complex flows.

Previous stability investigations of the jet in cross flow (Alves et al. 2007, 2008) have been based on various inviscid base flows adapted from the vortex sheet model of Coelho and Hunt (1989). They found that growth rates increase as the jet inflow ratio $R \equiv V_{\text{jet}}/U_\infty$ decreases. Recently, Megerian et al. (2007) found experimentally that for a low jet inflow ratio $R < 3.5$ external excitations have a small impact on the flow response, in contrast to the significant effect of forcing for larger values of R. This indicates a transition from a globally unstable flow in which intrinsic self-sustained global oscillations are present to a convectively unstable flow that exhibits a noise-amplifying behaviour (Huerre 2000).

As it has already been told, the phenomenon of a jet in a cross stream has been widely investigated for many years (Bagheri et al. 2009; Coelho and Hunt 1989; Keffer and Baines 1963; Chassing et al. 1974; Andreopoulos and Rodi 1984; Fric and Roshko 1994; Kelso and Smits 1995; Eiff and Keffer 1997; Blanchard et al. 1999; Yuan et al. 1999; Lim et al. 2001; Rivero et al. 2001; Su and Mungal 2004), and the special attention focused on the large-scale flow structure development, jet trajectories, scalar mixing and transport properties, and other associated flow phenomena. The interest in this phenomenon is driven partly by its immense applicability in aerodynamic flow control, combustion optimizations and jet-mixing enhancements, just to name a few. The highly intricate three-dimensional nature of the flow, coupled with the lack of complete understanding of the flow structures has provided further impetus for its study.

The deflection of a jet by the cross-flow causes a significant realignment of both the jet and cross-flow boundary layer vorticity fields, resulting in the formation of several major vortex systems about which it has been told above. In Fig. 1.1 a schematic of these major vortex systems for a typical round jet is shown. Some of these vortex structures have been found to be responsible for the significant entrainment of the cross-flow fluid by the jet as the latter penetrates into the former.

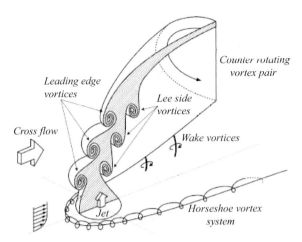

Fig. 1.1 A schematic drawing of large-scale vortex structures produced by a round jet exhausting normally into cross flow. The *shaded region* indicates the cross-section of the deflected jet cut along the jet center line in the cross flow direction (New et al. 2006)

For example, the near-field entrainment in the vicinity of the jet exit is influenced by the leading-edge and lee-side vortices formed along the jet/cross-flow interfaces (Coelho and Hunt 1989), while the far-field entrainment is dominated by the counter rotating vortex pair system further downstream (Smith and Mungal 1998). Recent studies by Haven and Kurosaka (1997), New et al. (2003), Plesniak and Cusano (2005) have found that changes to the jet geometries can have significant impacts on the resultant vortex structures, entrainment and jet penetration. On the other hand, Coelho and Hunt (1989) and Hasselbrink and Mungal (2001) noted that the entrainment process can affect the resultant jet penetration into the cross-flow, with higher near field entrainment causing a greater deflection of the transverse jet and vice versa. Hence, any changes to the generation of the leading-edge and lee-side vortices are likely to have an appreciable impact on the resulting flow field.

To date, most of the experimental and numerical investigations on a jet in cross flow are restricted to jets with the top-hat mean velocity profile at the nozzle exit. This is partly because most engineering applications involved high-speed flows or/and required the use of short entrance lengths for the jets. Also, imposing such an initial condition can simplify theoretical and computational analyses immensely. However, recent flow visualization study by New et al. (1999) has found that different jet velocity profiles namely, parabolic and top-hat, can also influence the development of near-field flow structures. More recently, Muppidi and Mahesh (2005) carried out DNS calculations on the effects of jet velocity profiles on the trajectories of a round jet in cross flow, and found significant deviations between the trajectories of a top-hat and parabolic round jet in cross flow. This leads them to propose an alternative scaling for the jet trajectories, by taking into consideration the relative inertia of the jet and the cross-flow. Hale et al. (2000), Peterson and Plesniak (2004) have also examined the issue of non-uniform jet exit velocity conditions, however, they did not specifically investigate top-hat and parabolic jet in cross flow.

Global instability of a round jet with a parabolic mean velocity profile at the nozzle exit is investigated numerically in the work (Bagheri et al. 2009) where global unstable eigenmodes of low and high frequencies are separated. The qualitative and quantitative data on development of a round jet both with top-hat and parabolic mean velocity profile at the nozzle exit are presented in the work by New et al. (2006). Controllable suppression of the instabilities or flow "relaminarization" at certain characteristics of the hydrodynamical system of a round jet in a cross flow evolution has been investigated by Watson and Sigurdson (2008). Turbulent round jet in a turbulent boundary layer was studied by Gopalan et al. (2004). As a whole, the last studies directed on the full understanding of the cross flow effect on a round jet with top–hat and parabolic mean velocity profile remain yet completely unclear and demand further investigations.

In what follows, the above and related aspects of the macro- and microjet instability and combustion are discussed according to the authors' experience in experimental work on the present topic. Research data obtained at low subsonic velocities on round and plane, free and near-wall jets of different scales are summarized to illustrate the dynamics of jet flows in different initial and environmental conditions.

Studying the round jet evolution depending on initial conditions at the nozzle exit and acoustic environment is very important for understanding the mechanisms of jet combustion. The combustion process is a chemical reaction; therefore, it has been studied for a long time within the framework of one of the areas of chemical science. However, combustion, for example, jet combustion also involves physical processes, such as heat and mass transfer, occurrence and development of coherent structures, etc.

It is natural that combustion should be considered in chemical and physical aspects. For example, one expects that the process of spontaneous ignition of a homogeneous gas mixture is a chemical reaction; however, various really observable processes of spontaneous ignition are also caused by physical phenomena. The major factors determining the character of combustion of diffusion flames are such physical processes as diffusion and mixing. The velocity of flame propagation in a homogeneous gas environment is determined by heat transfer and diffusion rather than by the chemical reaction rate (Kumagai 1979).

Flame motion over a gas mixture is referred to as flame propagation. Thus, the gas mixture is divided into two parts: burnt gas through which the flame has already passed and unburnt gas, which will soon enter the flame region. The border between these two parts of a burning gas mixture is referred to as *the flame front*. Flame propagation can be of two types: a detonation wave and a combustion wave. The combustion wave is characterized by the fact that the flame propagates by means of heat transfer and diffusion of active molecules from the flame front, consistently transforming the unburnt gas to products of combustion. The combustion wave propagation velocity is much lower than the speed of sound, and the detonation wave velocity exceeds the speed of sound. The burning rate is defined as the normal (to the flame front) velocity component of flame propagation over the (initial) gas mixture that has not burned yet. The burning rate differs thus from the speed of spatial motion of the flame, i.e., from the velocity of flame propagation. The flame front motion is caused by the motion of the stream of the gas mixture and expansion of the combustion products. Therefore, the flame propagation velocity is the sum of velocities of these streams and the burning rate (Kumagai 1979).

If the fuel is sprayed in air or oxygen and mixed with them during burning, the flame refers to diffusion flames. Jet combustion in the flow of a gas through a nozzle into the open space or a chamber is a widespread kind of burning. Two types of combustion are distinguished: combustion of a premixed fuel-air mixture and combustion of a non-premixed fuel-air mixture, where the combustion process is defined by diffusion of two separate streams of fuel and oxidizer.

Diffusion is a determining factor that affects the evolution of the characteristics of the diffusion flame as compared to the same characteristics of the flame during combustion of a premixed fuel-air mixture. The diffusion flame can be stationary (in the case of continuous combustion in various burners and gas turbines) and non-stationary (intermittent combustion, for example, in the internal combustion engine). The most typical diffusion jet flame is formed in the case of ignition of a jet of a combustible gas flowing from a long tube of small diameter into the air of the atmosphere. When the jet speed is insignificant, the flow, naturally, is laminar, the flame border is steady, the flame shape is smooth, and combustion proceeds quietly. As the jet velocity increases, the flame height grows. However, such a picture is observed only up to some limit of the jet velocity. With a further increase in the jet velocity, the flame border becomes unstable, and instability at the beginning arises only at the top of the flame, and then it gradually propagates downward to the nozzle exit. The flame height simultaneously sharply decreases. With a subsequent increase in the jet velocity, the flame height ceases to depend on the jet velocity and remains approximately constant, the flame border sharply oscillates, and combustion is accompanied by intense noise. As the jet velocity further increases, the flame is lifted from the nozzle exit and is stabilized at some distance above it. Flames of such type refer to lifted flames. The jet region in which the flame height is almost independent of the jet velocity corresponds to a turbulent jet flow. It is also the so-called turbulent diffusion jet flames. The jet region in which the flame height grows as the jet velocity increases refers to laminar diffusion flames. Between them there is some transition area. The transition from a laminar diffusion flame to a turbulent flame is defined by a change in the jet flow characteristics. The turbulent flame propagation velocity depends on the flow velocity, and also on both the degree and scale of turbulence. Besides, temperature changes due to combustion naturally affect the jet flow. Therefore, the jet flow in which there is no combustion differs from the jet flow with combustion (Kumagai 1979). The dependence of the diffusion flame length on the jet flow velocity is shown in Fig. 1.2, taken from the work (Vulis et al. 1968).

Numerous analytical, numerical, and experimental works (Vanquickenborne and van Tiggelen 1966; Peters and Williams 1983; Byggstoyl and Magnussen 1983; Schefer et al. 1994) are devoted to studying the diffusion combustion process (in

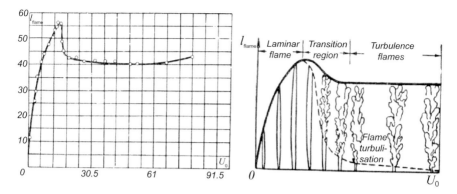

Fig. 1.2 Dependence of the diffusion flame length on the nozzle gas flow velocity (Vulis et al. 1968)

particular, jet combustion). For example, lifted turbulent flames cause significant interest because they include many fundamental mechanisms of flame stabilization control and flame attenuation in practical burners. The authors of the work (Vanquickenborne and van Tiggelen 1966) believe that the fuel and air in lifted flames completely preliminary mix up before ignition and that flame stabilization occurs when the local flow velocity along the stoichiometry contour is equal to the turbulent flame propagation velocity. The authors of (Peters and Williams 1983) give reasons why the fuel and air do not mix up at a point of flame stabilization and why flame lifting can be explained under conditions of the concept of a laminar flame. The authors of the model (Byggstoyl and Magnussen 1983) believe that the lifted flame is stabilized when the fuel and air are mixed at a molecular level near the stoichiometry conditions [mass (volume) ratio of reactants] and the local jet flow velocity is equal to the maximum flame propagation velocity. Which of these models is closer to the true physical picture remains unclear. Experimental studies carried out with a purpose of defining the acceptability of these theoretical models (Schefer et al. 1994) showed that local stoichiometry and characteristics of turbulence instead of dissipation are the initial factors for flame stability control. In the work (Polejaev 2010), it is shown that the turbulent combustion velocity repeatedly exceeds the laminar combustion velocity. A large role in the mechanism of flame stabilization belongs to coherent structures of the turbulent jet. There are significant interactions between the large-scale structure connected with the central jet and the flame region.

In the experiments (Muniz and Mungal 1997), it is revealed that the mean lift-off height of the turbulent diffusion flame grows with an increase in the accompanying air flow velocity and the jet velocity at the nozzle exit. Fluctuations of the flame lift-off height tend to increase with the growth of both the flame lift-off height and the jet width. This observation shows that the flame attenuation mechanism can be connected with the large-scale vortex structure. Two criteria of flame stabilization were found (Muniz and Mungal 1997). The first criterion defines that the flame is stabilized by itself when the local gas velocity is close to the velocity of propagation of the premixed laminar flame and does not exceed $3SL$, where SL is the maximum velocity of premixed laminar flame propagation. This result explains why small velocities of the accompanying flow (entrained air) can have such a big effect on flame stabilization and suggests that the flame decays if the accompanying flow velocity exceeds this value. The second criterion of flame stabilization says that the structure of fuel/air mixing should be within flammability limits. It is desirable to be close to stoichiometry conditions where the maximum flame propagation velocity is reached. The authors of the work (Everest et al. 1996) show that the combustibility region of reagents is located on the base (flame leading edge) of the lifted flame and that premixed flame propagation can explain the upstream motion of the flame base. The thickness of this combustibility layer changes as a function of the mixing characteristics of the large-scale vortices generated in the shear layer.

An important factor during combustion is the influence of acoustics. Ordered vortex structures (Huerre and Monkewitz 1990) are strongly susceptible to acoustic disturbances. Hence, forced disturbances can be used to change significantly and even control the development of jet structures (Crow and Champagne 1971), for example, for turbulence reduction (Zaman and Hussain 1981; Hussain and Hasan 1985). On the other hand, acoustic fields are usually present in the combustion process; their occurrence creates new conditions for the jet flame environment. Knowing the nature of these fields, it is possible to help remove some complications of combustion. Thus, various authors studied diffusion or premixed flames in the presence of acoustics excitation. In the case of laminar diffusion flames, acoustics is used in the work (Hertzberg 1996) in which particular importance is given to the lifted flame behavior. In lifted jet flames, acoustics can be used to change the process of their combustion for reduction of soot and NO_x emissions (Chao et al. 1996). In the work (Chao et al. 1994), it is shown that the lifted flame height is reduced by jet amplification with its longitudinal excitation at the natural frequency of the jet, but the importance of three-dimensional instabilities is not emphasized. In the work (Suzuki et al. 2011), the flame bifurcation process is investigated for combustion of round and plane fuel jets under the influence of a transverse acoustic field.

The purpose of this part of our work is an experimental study of the influence of initial conditions at the nozzle exit and the acoustic effect on the structure and characteristics of the development of laminar and turbulent macro- and microjets with and without combustion. Preliminary studies of laminar and turbulent propane macrojet diffusion combustion without the acoustic effect are conducted to find the influence of changes in the initial conditions at the nozzle exit on the combustion process. Diffusion and premixed combustion of propane round and plane microjets in a transverse acoustic field is experimentally studied.

References

Abid M (1993) Simulation numeriques directes de la dynamique de transition tridimensionnelle des jets axisymetriques. PhD. Thesis, Ecole Normale Superieure de Paris

Alves L, Kelly R, Karagozian A (2007) Local stability analysis of an inviscid transverse jet. J Fluid Mech 581:401–418

Alves L, Kelly R, Karagozian A (2008) Transverse-jet shear-layer instabilities. Part 2. Linear analysis for large jet-to-crossflow velocity ratio. J Fluid Mech 602:383–401

Andreopoulos J, Rodi W (1984) Experimental investigation of jets in a crossflow. J Fluid Mech 138:93–127

Bagheri S, Schlatter Ph, Schmid PJ, Henningson DS (2009) Global stability of jet in crossflow. J Fluid Mech 624:33–44

Bernal LP, Roshko A (1986) Streamwise vortex structure in plane mixing layers. J Fluid Mech 170:499–519

Blanchard JN, Brunet Y, Merlen A (1999) Influence of a counter rotating vortex pair on the stability of a jet in a cross-flow: an experimental study by flow visualizations. Exp Fluids 26:63–74

Brancher P, Chomaz JM, Huerre P (1994) Direct numerical simulation of round jets: vortex induction and side jets. Phys Fluids 6:1768–1775

Byggstoyl S, Magnussen BF (1983) A model for flame extinction in turbulent flow. In: Bradbury LJS, Durst F, Schmidt FW, Whitelaw JH (eds) Fourth Symposium on Turbulent Shear Flows Karlsruhe, pp 10.32–10.38

Chao YC, Yuan T, Jong YC (1994) Measurement of the stabilization zone of a lifted jet flame under acoustic excitation. Exp Fluids 17:381

Chao YC, Yuan T, Tseng CS (1996) Effects of flame lifting and acoustic excitation on the reduction of NOx emissions. Combust Sci Technol 113:49–60

Chassing P, George J, Claria A, Sananes F (1974) Physical characteristics of subsonic jets in a cross-stream. J Fluid Mech 62:41–64

Coelho S, Hunt J (1989) The dynamics of the near field of strong jets in crossflows. J Fluid Mech 200:95–120

Crow SC, Champagne FH (1971) Orderly structure in jet turbulent. J Fluid Mech 48:547–591

Demare D, Baillot F (2001) The Role of secondary instabilities in the stabilization of a nonpremixed lifted jet flame. Phys Fluids 13:2662–2669

Eiff OS, Keffer JF (1997) On the structures in the near-wake region of an elevated turbulent jet in a crossflow. J Fluid Mech 333:161–195

Everest D, Feikema D and Driscoll JF (1996) A study of the mechanism of jet flame liftoff—Based on images of the strained flammable layer. Proc Combust Inst 26:129 (Twenty Sixth Symposium (International) on Combustion, The Combustion Institute Pittsburgh)

Fric TF, Roshko A (1994) Vortical structure in the wake of a transverse jet. J Fluid Mech 279:1–47

Gopalan S, Abraham BM, Katz J (2004) The structure of a jet in cross flow at low velocity ratios. Phys Fluids 16(6):2067–2087

Guinevskiy AS, Vlasov EV, Karavosov RK (2001) Acoustic control of turbulent jets. Nauka, Moscow (in Russian)

Hale CA, Plesniak MW, Ramadhyani S (2000) Structural features and surface heat transfer associated with a row of short-hole jets in crossflow. Int J Heat Fluid Flow 21:542–553

Hasselbrink EF, Mungal MG (2001) Transverse jets and jet flames. Part 1. Scaling laws for strong transverse jets. J Fluid Mech 443:1–25

Haven BA, Kurosaka M (1997) Kidney and anti-kidney vortices in crossflow jets. J Fluid Mech 352:27–64

Hertzberg JR (1996) Conditions for a split diffusion flame. Combust Flame 109:49

Hu H, Saga T, Kobayashi T, Taniguchi N, Liu H, Wu S (1999) Research on the rectangular lobed exhaust ejector/mixer systems. Trans Jpn Soc Aeronaut Space Sci 41:187–197

Hu H, Saga T, Kobayashi T, Taniguchi N (2001) A study on a lobed jet mixing flow by using stereoscopic particle image velocimetry technique. Phys Fluids 13:3425–3441

Huerre P, Monkewitz PA (1990) Local and global instabilities in spatially developing flows. Annu Rev Fluid Mech 22:473

Huerre P (2000) Open shear flow instabilities. In: Batchelor GK, Mofatt HK, Worster MG (eds) Perspectives in fluid dynamics. Cambridge University Press, Cambridge, pp 159–229

Hussain AKM, Hasan MAZ (1985) Turbulent suppression in free turbulent shear flows under controlled excitation. Part 2. J Fluid Mech 150:159

Keffer JF, Baines WD (1963) The round turbulent jet in a cross wind. J Fluid Mech 15:481–496

Kelso RM, Smits AJ (1995) Horseshoe vortex systems resulting from the interaction between a laminar boundary-layer and a transverse jet. Phys Fluids 7:153–158

Kelso R, Lim T, Perry A (1996) An experimental study of round jets in cross-flow. J Fluid Mech 306:111–144

Kuchar AP, Chamberlin R (1980) Scale model performance test investigation of exhaust system mixers for an energy efficient engine (E3). AIAA Paper No 80-0229

Kumagai S (1979) Combustion. Translation with Japanese, Publishing house "Chemistry", Moscow, pp 1–256 (In Russian)

Lasheras JC, Cho JS, Maxworthy T (1986) On the origin and evolution of streamwise vortical structures in plane free shear layer. J Fluid Mech 172:231–247

Liepman D, Gharib M (1992) The role of streamwise vorticity in the near-field of round jet. J Fluid Mech 245:643–668

Lim TT, New TH, Luo SC (2001) On the development of large scale structures of a jet normal to a cross-flow. Phys Fluids 3:770–775

Megerian S, Davitian L, Alves L, Karagozian A (2007) Transverse-jet shear-layer instabilities. Part 1. Experimental studies. J Fluid Mech 593:93–129

Metcalfe RW, Orszay SA, Brachet ME, Menon S, Riley JJ (1987) Secondary instability of a temporally growing mixing layer. J Fluid Mech 184:207–219

McCormic DC, Bennett JC (1994) Vortical and turbulent structure of a lobed mixer free shear layer. AIAA J 32:1852–1856

Monkewitz PA, Lehmann B, Barsikow B, Bechert DW (1989) The spreading of self-excited hot jets by side jets. Phys Fluids 1:446–456

Monkewitz PA, Pfizenmaier E (1991) Mixing by side jets in strongly forced and self-excited round jets. Phys Fluids 3:1356–1364

Muniz L, Mungal MG (1997) Instantaneous flame-stabilization velocities in lifted jet diffusion flames. Combust Flame 111(1/2):16–31

Muppidi S, Mahesh K (2005) Study of trajectories of jets in crossflow using direct numerical simulations. J Fluid Mech 530:81–100

Muppidi S, Mahesh K (2007) Direct numerical simulation of round turbulent jets in crossflow. J Fluid Mech 574:59–84

New TH, Lim TT, Luo SC (1999) On the effects of velocity profiles on the topological structures of a jet in cross-flow. Proc TSFP 1:647–652

New TH, Lim TT, Luo SC (2003) Elliptic jets in cross-flow. J Fluid Mech 494:119–140

New TH, Lim TT, Luo SC (2006) Effects of jet velocity profiles on a round jet in cross-flow. Exp Fluids 40(3):859–875

Paterson RW (1984) Turbofan forced mixer nozzle flow field: a benchmark experimental study. ASME J Eng Gas Turbines Power 106:692–700

Peters N, Williams F (1983) Liftoff characteristics of turbulent jet diffusion flames. AIAA J 21 (1):423–429

Peterson SD, Plesniak MW (2004) Evolution of jets emanating from short holes into crossflow. J Fluid Mech 503:57–91

Plesniak MW, Cusano DM (2005) Scalar mixing in a confined rectangular jet in crossflow. J Fluid Mech 524:1–45

Poletaev YuV (2010) About turbulent jets and physics of the gas jet-flame combustion. In: Proceedings of the International Conference. Nonlinear problems of the hydrodynamical stability theory and turbulence, Moscow, pp 137–144 (in Russian)

Power GD, McClure MD, Vinh D (1994) Advanced IR suppresser design using a combined CFD/Test ApproaChap. AIAA Paper No 94-3215

Presz WM, Reynolds G, McCormic D (1994) Thrust augmentation using mixer-ejector-diffuser system. AIAA Paper No 94-0020

Rivero A, Ferre JA, Giralt F (2001) Organized motions in a jet in crossflow. J Fluid Mech 444:117–149

Schefer RW, Namazian M, Kelly J (1994) Stabilization of lifted turbulent-jet flames. Combust Flame 99(1):75–86

Smith LL, Majamak AJ, Lam IT, Delabroy O, Karagozian AR, Marble FE, Smith OI (1997) Mixing enhancement in a lobed injector. Phys Fluids 9:667–672

Smith SH, Mungal MG (1998) Mixing, structure and scaling of the jet in crossflow. J Fluid Mech 357:83–122

Su LK, Mungal MG (2004) Simultaneous measurements of scalar and velocity field evolution in turbulent crossflowing jets. J Fluid Mech 513:1–45

Suzuki M, Ikura S, Masuda W (2011) Comparison between acoustically-excited diffusion flames of tube and slit burners. In: Proceedings of the 11th Asian Symposium on Visualization, Niigata, Japan, pp 1–6

Vanquickenborne L, van Tiggelen A (1966) The stabilization mechanism of lifted diffusion flames. Combust Flame 10(1):59–69

Vulis LA, Ershin ShA, Yarin LP (1968) The theory of a gas flame. Leningrad Publishing house "Energy", pp 1–203 (in Russian)

Watson GMG, Sigurdson LW (2008) The controlled relaminarization of flow velocity ratio elevated jet-in-crossflow. Phys Fluids 20:094108(1)–094108(15)

Yuan LL, Street RL, Ferziger JH (1999) Large-eddy simulations of a round jet in crossflow. J Fluid Mech 379:71–104

Zaman KBMQ (1997) Axis switching and spreading of an asymmetric jet: the role of coherent structure dynamics. J Fluid Mech 316:1–20

Zaman KBMQ, Hussain AKMF (1981) Turbulence suppression in free shear flows by controlled excitation. J Fluid Mech 103:133–145

Chapter 2
Evolution and Breakdown of a Subsonic Round Jet

We begin with the dynamics of a round jet emphasizing the contribution of longitudinal disturbances to the perturbed flow pattern. Important issues are generation, spatial development, and interaction of the three-dimensional structures with the ring vortices. Supplementary material to this chapter is given by the multimedia files titled Chap. 2: "Multimedia files Nos. 2.1–2.8" (http://extras.springer.com).

2.1 Experimental Technique

A test facility used for modeling of the round jet was that with a Vitoshinsky nozzle, a settling chamber (1) and deturbulizing grids (2) as is shown in Fig. 2.1. The air stream in the channel is generated by a fan. In most of the experimental runs which were performed, the jet core velocity U_0 at the nozzle exit of the diameter $d = 40$ mm was equal to 4 m/s making the Reynolds number Re $= U_0 d/v = 10{,}600$. Experimental results presented in what follows were obtained by flow visualization and hot-wire measurements.

At the visualization, smoke was injected into the jet from the fan side. The flow patterns were recorded by a digital camera (9) providing a general view of the jet and its slices both in streamwise and cross sections through illumination of the flow by a laser sheet of 0.5 mm thickness (7). Also, the flow was illuminated by a light sheet of a filament lamp using a narrow slot of a diaphragm. To examine the effects of acoustic excitation, the jet was forced by a dynamic loudspeaker (8) generating harmonic oscillations at controlled variation of their frequency and sound pressure level. The acoustic forcing was also used to synchronize the rolling up of the Kelvin–Helmholtz ring vortices with the laser illumination periodically switched on for short time intervals to record stroboscopic patterns of the interaction between the ring vortices and the three-dimensional perturbations.

© The Author(s) 2016
V.V. Kozlov et al., *Visualization of Conventional and Combusting Subsonic Jet Instabilities*, SpringerBriefs in Applied Sciences and Technology, DOI 10.1007/978-3-319-26958-0_2

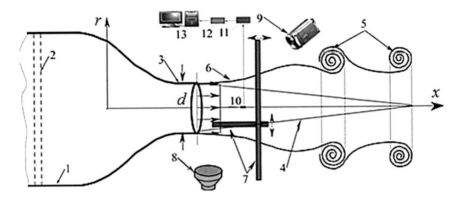

Fig. 2.1 Experimental set-up: settling chamber (*1*), deturbulizing grids (*2*), nozzle (*3*), potential core of the jet (*4*), Kelvin–Helmholtz vortices (*5*), 3D disturbances (*6*), positions of the laser sheet (*7*), loudspeaker (*8*), digital camera (*9*), hot-wire probe (*10*), hot-wire anemometer (*11*), ADC (*12*), PC (*13*)

The hot-wire measurements of the streamwise mean velocity and fluctuations were performed with a DISA constant-temperature anemometer (11) and a one-wire probe of 1-mm length and 5-μm diameter made of gilded tungsten (10). The probe was operated at 80 % overheat ratio and calibrated through the modified King's law

$$U = k_1 \left(E^2 - E_0^2\right)^{1/n} + k_2 (E - E_0)^{1/2},$$

where E and E_0 are the output voltages of the anemometer at flow velocity switched on and off, respectively, and k_1, k_2, n are constants. The index $1/n$ is usually about 2, the constant k_2 allows for free convection near the wall where the flow velocity is close to zero. The maximum error at the probe calibration did not exceed 1 % of U_0. The hot-wire signal was digitized by a 16-bit analog-to-digital converter (12) for further acquisition and processing with a personal computer (13). The same technique applies to hot-wire results on other configurations of jet flows discussed in the following chapters.

2.2 Velocity Characteristics

Distributions of the mean and fluctuation velocity components at different streamwise distances from the nozzle exit are shown in Fig. 2.2. The jet was traversed by the hot-wire probe in four cross sections at $x/d = 0.2, 0.3, 0.55$, and 0.8 where x is the streamwise coordinate measured from the nozzle exit. In the figure, the radial distance r is normalized by the channel radius $d/2$, the mean flow velocity U and rms amplitude of disturbances u' are reduced by the maximum value U_0 at the jet axis. The mean-flow profiles demonstrate a wide jet core and a narrow region

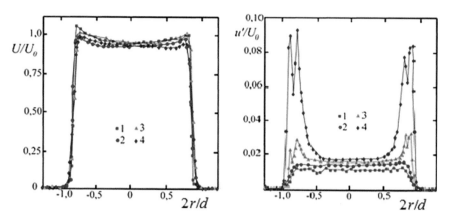

Fig. 2.2 Radial distributions of the mean (*left*) and fluctuation (*right*) streamwise velocity components at different distances from the nozzle exit: $x/d = 0.2$ (*1*), 0.3 (*2*), 0.55 (*3*), and 0.8 (*4*); $U_0 = 4$ m/s

of strong velocity gradient at the jet periphery. Virtually, such distributions called "top-hat" profiles result in the generation of the ring vortices (Kelvin–Helmholtz ones) close the nozzle exit. A similar phenomenon is observed at the blowing of vortex rings by a tricky smoker. To do this, he draws an outward breath through the lips rounded off in the form of a nozzle and creates the velocity profile of the jet, similar to that presented in Fig. 2.2.

2.3   Interaction of the Primary Vortices with the Longitudinal Disturbances

An example of smoke visualization of the laminar jet breakdown at Re = 10,000 is given by Van Dyke (1986) where one can observe the Kelvin–Helmholtz instability at the jet periphery. Then, this flow region rolls up into vortex rings and after that the jet suddenly becomes turbulent. The smoke ring looks like a tightly twisted toroidal helix appearing due to the roll up of the vortex sheet that comes off the nozzle edge. The transition of laminar jet to the turbulent state is primarily caused by the Kelvin–Helmholtz instability, and further, by the secondary instability of vortex rings. The latter, often called the Widnall instability, is also illustrated by Van Dyke (1986) and appears as growing waves around the vortex ring. The origination of these waves is usually explained by the interaction of the Kelvin_Helmholtz and the streamwise vortices produced using a specially profiled nozzle as well as appearing due to the secondary instability of the jet. The interaction of the longitudinal vortices generated by the jet secondary instabilities with the ring vortices is investigated in detail by Demare and Baillot (2001). They have shown that between the ring vortices there is an emission of longitudinal

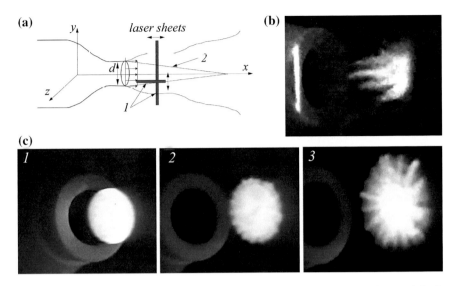

Fig. 2.3 Natural longitudinal disturbances in the near field of the round jet: scheme of the jet illumination by thin light beams marked by *1* and *2* (**a**); flow pattern in the streamwise section at $r/d = 0.4$ (**b**); the same in cross sections at $x/d = 0.4$ (*1*), 0.68 (*2*), and 0.86 (*3*) (**c**); $U_0 = 4$ m/s (see Presentation Chap. 2: "Multimedia files Nos. 2.1–2.4") (http://extras.springer.com)

counter-rotating vortices in the azimuthal or radial direction. This leads to enhancement of the jet mixing with the surrounding air over the entire periphery of the jet.

The three-dimensional disturbances which are localized at the jet periphery and arise in natural conditions close to the nozzle exit are visualized using a light sheet as is shown in Fig. 2.3. In the present case, one expects several reasons for their origination including surface roughness, presence of grids and honeycombs in the settling chamber, etc. The streamwise evolution of the generated perturbations and their interaction with the ring vortices result in the jet turbulization.

Naturally occurring streamwise elongated perturbations are subject to radial oscillations complicating the observation of flow details. Therefore, they were modeled using controlled roughness elements, each of 0.2-mm height and 5-mm width, glued on the internal surface of the nozzle near its exit at their spacing correlating to the azimuthal scale of the natural perturbations. In this way, the development and interactions of the three-dimensional disturbances could be stabilized.

Visualization results in the form of stroboscopic patterns obtained with the laser sheet illumination illustrate the jet slices at different distances from the nozzle in Fig. 2.4. At the nozzle exit (Fig. 2.4*b*), the jet boundary is nearly sinusoidal along the full circumference. At $x/d = 0.3$ (Fig. 2.4*c*) the sinusoidal contour is distorted by the azimuthal beams as a result of the interaction of the first ring vortex with the longitudinal perturbations generated at the roughness elements. Then, at $x/d = 0.5$

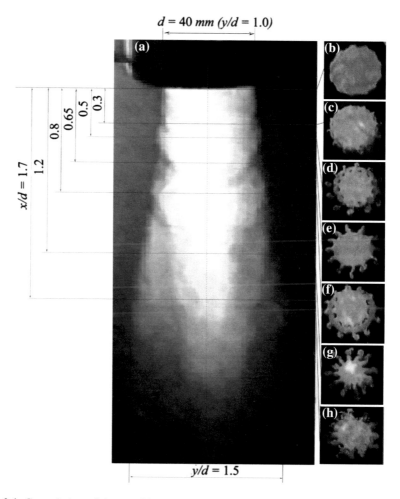

Fig. 2.4 General view of the round jet (**a**) and its cross sections at different distances from the nozzle exit (**b–h**) (see Presentation Chap. 2: "Multimedia files Nos. 2.5–2.8") (http://extras.springer.com)

(Fig. 2.4*d*), the distortion increases in the center of the first ring vortex, and the beams take the form of thin pins with the mushroom-like structures arising at their tips, similar to those described by Demare and Baillot (2001). These are identified as a pair of vortices rotating in opposite directions. In the region between the first and second ring vortices at $x/d = 0.65$ (Fig. 2.4*e*), the ring vortices are not observed, but the intensity of beams increases. At $x/d = 0.8$ (Fig. 2.4*f*), some dynamics of the beams is found, especially after the passage of the second ring vortex at $x/d = 1.2$ (Fig. 2.4*g*). Then, while moving downstream, the intensive mixing of the jet with the surrounding air occurs through the development of the beams and, finally, the entire jet becomes almost turbulent at $x/d = 1.7$ (Fig. 2.4*h*).

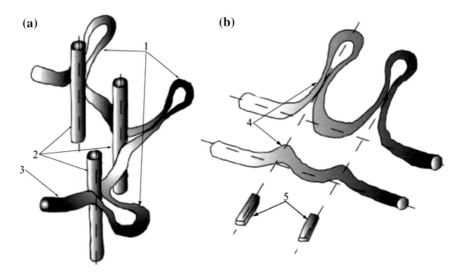

Fig. 2.5 Three-dimensional deformation of the ring vortex (**a**) as compared to that of the Tollmien–Schlichting wave in a boundary layer (**b**): Λ- or Ω-like vortices (*1*), longitudinal disturbances (*2*), ring vortex (*3*), 3D distortions of the 2D instability wave (*4*), roughness elements (*5*)

The above process of the laminar jet breakdown associated with the interaction of the three-dimensional disturbances with the ring vortices resembles the deformation of two-dimensional instability waves in boundary layers which is followed by generation of so-called Λ- or Ω-like structures. In a boundary layer they appear as two vortices rotating in opposite directions with a "head" at their tips. In the classical experiment by Klebanoff et al. (1962) the Λ-structures were observed in a flat-plate flow at a distortion of the Tollmien–Schlichting wave by local boundary-layer non-uniformities. In the present case, such non-uniformities are the longitudinal perturbations as is sketched in Fig. 2.5. A two-dimensional ring vortex, while passing through the flow region perturbed by the longitudinal disturbances, interacts with them and undergoes a three-dimensional deformation which results in the characteristic bursts well seen all over the jet periphery. Those observed in Fig. 2.4 are likely to be the "heads" of Λ- or Ω-like structures.

Note that in the boundary layer the Λ-structures start growing from the wall at local flow velocity close to zero towards the edge of the layer where the velocity tends to its external-flow value. Vice versa, in the jet two counter rotating vortices are ejected from the high-speed region of the vortex ring to surrounding air at rest so that the Λ-structure is elongated at its downstream propagation. Flow patterns taken in the region of the "heads" of Λ- or Ω-like structures (at the radial distance R_1) and at their origination from the ring vortex (R_2) are presented in Fig. 2.6. Three smoke streaks are observed in the region of the "heads" (Fig. 2.6*c*). As is distinctly seen in Fig. 2.6*d*, their length is larger than the distance between two adjacent ring

Fig. 2.6 General view of the round jet (**a**); a cross section (**b**); streamwise sections at R_1 (**c**) and R_2 (**d**); positions of the ring vortices and the longitudinal perturbations are marked by *1* and *2*, respectively (see Presentation Chap. 2: "Multimedia files Nos. 2.5–2.7, 2.8") (http://extras.springer.com)

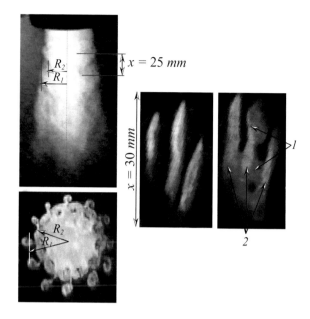

vortices. Thus the streamwise elongated perturbations are likely spreading far downstream up to the jet turbulization.

2.4   High-Frequency Instability of the Longitudinal Perturbations

Inducing local velocity gradients in shear layers, the three-dimensional disturbances are often prone to high-frequency oscillations stimulating the laminar-turbulent transition. One expects that a similar phenomenon may have a valuable contribution to the round jet turbulization, as well. To clarify this point, an oscillatory disturbance was injected by periodic blowing/suction of air through a tiny hole at the nozzle surface near one of the roughness elements (Fig. 2.7). As a result, the three-dimensional flow perturbation associated with the element became more pronounced than is seen in the top of Fig. 2.7*b* (*1*). Further downstream, the neighboring longitudinal structures are also intensified being involved in the jet turbulization, see Fig. 2.7*b* (*2*). These observations indicate that the laminar jet breakdown can be controlled through the excitation of secondary instability of the streamwise elongated disturbances.

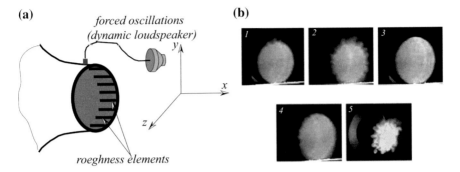

Fig. 2.7 Secondary instability of the longitudinal structures: experimental scheme (**a**); cross sections of the jet at $x/d = 0.2$ (*1* and *3*), 0.4 (*2* and *4*), and 0.68 (*5*); flow patterns *1*, *2*, and *5* are affected by the secondary disturbances while images *3* and *4* were taken in the absence of the forced oscillations (**b**); $U_0 = 4$ m/s (see Presentation Chap. 2: "Multimedia file No. 2.4") (http://extras.springer.com)

2.5 Jet Dynamics Under Acoustic Excitation

Two effects of the external acoustic forcing which can be involved in control of jet flows are just mentioned here. One of them is modification of the scale of the ring vortices. An example is given in Fig. 2.8 where the perturbed flow patterns visualized at variation of the excitation frequency can be compared.

Another effect to be emphasized is acoustic stimulation of the formation of longitudinal perturbations and their interaction with the ring vortices. Figure 2.9 shows that under the excitation, the three-dimensional deformations of the jet periphery become obviously stronger taking a mushroom-like shape. Thus, the acoustic forcing ensures more profound mixing of the jet with the surrounding air.

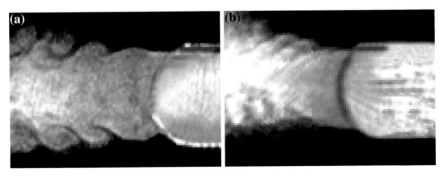

Fig. 2.8 Acoustic effect on the ring vortices: excitation frequencies $f = 110$ Hz (**a**) and $f = 250$ Hz (**b**); $U_0 = 5$ m/s

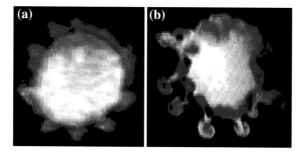

Fig. 2.9 Visualization of the jet cross section $x/d = 0.5$ without acoustic oscillations (**a**) and under the excitation at 140-Hz frequency and sound pressure level of 90 dB (**b**)

2.6 Effect of Flow Velocity on the Jet Evolution

Some flow patterns obtained in one and the same cross section of the jet are presented in Fig. 2.10 where one can trace the laminar flow breakdown at a fixed distance from the nozzle when increasing the flow velocity. The initial development of the three-dimensional perturbations (Fig. 2.10*a*, *b*), their interaction with the ring vortices resulting in the 3D bursts (Fig. 2.10*c–f*), and final transition to turbulence (Figs. 2.10*g*, *h*) are distinctly seen.

Finally, as criteria of the laminar and turbulent flow regimes the following can be taken. First, the hot-wire data on velocity perturbations at different streamwise stations (Fig. 2.2) show that the disturbances of the jet rapidly grow at $x = 32$ mm (approximately four times comparing to their intensity at $x = 8$, 12, and 22 mm) as an indication of a transition to the turbulent state. Then, a chaotic motion typical of

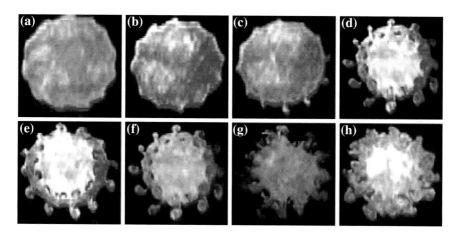

Fig. 2.10 Visualization of the jet cross section $x/d = 0.3$ at variation of flow velocity from $U_0 = 6$ to 13 m/s with the step of 1 m/s (from **a** to **h**) (see Presentation Chap. 2: "Multimedia files Nos. 2.6, 2.8") (http://extras.springer.com)

turbulent flow is seen in Figs. 2.4h and 2.10h, whereas in the upstream sections (Fig. 2.4b–g) and at a lower velocity of the jet core (Fig. 2.10a–f) an ordered flow structure is found that is typical of laminar flows and those at the early transitional stage.

Key points

Summarizing the above experimental results, we focus on the following aspects of the laminar jet breakdown:

- The longitudinal structures of velocity perturbations contributing to the laminar flow breakdown can be generated close to the jet origin, i.e., the nozzle exit.
- Interaction of the ring vortices with the longitudinal disturbances is similar to the deformation of two-dimensional instability waves in a boundary layer by local flow non-uniformities.
- A result of the interaction is the generation of "beams" in the form of Λ- or Ω-like structures spaced over the ring vortex.
- An intensive mixing of the jet with the surrounding air occurs in the region of the heads of Λ- or Ω-like structures enhancing the jet spreading and its transition to the turbulent state.
- Under the external acoustic forcing, the passage frequency and the scales of the ring vortices are modified as well as mixing of the jet with ambient air becomes more profound.

References

Demare D, Baillot F (2001) The role of secondary instabilities in the stabilization of a nonpremixed lifted jet flame. Phys Fluids 13:2662–2669
Klebanoff PS, Tidstrom KD, Sargent LM (1962) The three-dimensional nature of boundary layer instability. J Fluid Mech 12:1–34
Van Dyke M (1986) Album fluid motion. Mir, Moscow, in Russian

Chapter 3
Instability of Free and Wall Plane Jets

In this chapter we proceed with plane jets focusing, again, on the formation and development of longitudinal structures of laminar flow disturbances and their interaction with the Kelvin–Helmholtz vortices. Some visualization results on a free jet are presented and, then, a perturbed wall jet is considered in more detail. For the corresponding supplementary material see multimedia files titled Chap. 3: "Multimedia files Nos. 3.1–3.13" (http://extras.springer.com).

3.1 Three-Dimensional Perturbations of the Free Plane Jet

3.1.1 Experimental Set-up

The free plane jet emanating from a rectangular nozzle exit of 10×200 mm ($h \times l$) was examined as is sketched in Fig. 3.1. The streamwise elongated disturbances were generated in a controlled manner by roughness elements of 0.2-mm height, 20-mm length, and 5-mm width periodically mounted at 10 mm from each other at one side of the nozzle near its exit. The particular roughness spacing was chosen to produce the structures with their transverse scale of several shear layer thickness at which they naturally develop in near-wall boundary layers (Westin et al. 1994). Natural frequency of the Kelvin–Helmholtz vortices was stabilized by acoustic excitation of the jet using a loudspeaker which was placed at 200 mm from the nozzle and radiated sound waves at the frequency $f = 140$ Hz normally to the jet plane.

The jet was visualized in its cross-sections in the range of 5–85 mm from the nozzle, and also in streamwise sections of the shear layers by a laser sheet of about 0.3-mm thickness. The accuracy of the traversing system positioning the laser sheet was 0.1 mm. The laser was driven by squire wave signals at the sound frequency varying their relative pulse duration by means of a phase shifter so that the most interesting stages of the periodic phenomena could be captured. The visualization data were acquired as 25 frames per second. The stroboscopic images of the flow sections were recorded by a digital camera placed at an angle of 45° to the section plane at a distance of 0.5 m from the jet axis. During the experimental runs, the jet

© The Author(s) 2016
V.V. Kozlov et al., *Visualization of Conventional and Combusting
Subsonic Jet Instabilities*, SpringerBriefs in Applied Sciences and Technology,
DOI 10.1007/978-3-319-26958-0_3

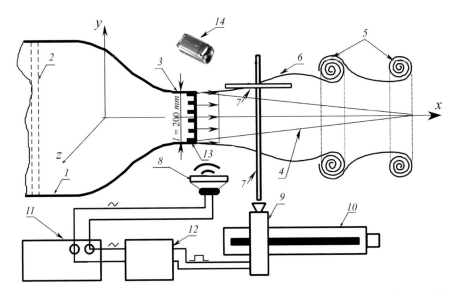

Fig. 3.1 Experimental set-up and the jet flow: settling chamber (*1*); grids (*2*); nozzle (*3*); potential jet core (*4*); Kelvin–Helmholtz vortices (*5*); shear layer (*6*); laser sheet (*7*); loudspeaker (*8*); laser (*9*); traversing mechanism (*10*); sound generator (*11*); phase shifter (*12*); roughness elements (*13*); video camera (*14*)

core velocity U_0 was in the range from 2 to 12.2 m/s corresponding to the Reynolds number Re $= hU_0/v$ varying from 1300 to 8100. Root-mean-square amplitude of background velocity perturbations near the nozzle exit at the jet axis and in the shear layer was not higher than 0.4 and 1.5 % of U_0, respectively.

3.1.2 Visualization of the Free Plane Jet

As is seen in Fig. 3.2, the Kelvin–Helmholtz vortices evolve in the shear layer together with its transverse modulation appearing as streamwise beams. Interacting with each other they result in a complex spatio-temporal periodic velocity perturbation.

Some images of the jet cross sections are given in Fig. 3.3. They demonstrate a result of the interaction between the longitudinal perturbations and the Kelvin–Helmholtz vortices, i.e., the development of periodic mushroom-like structures spreading from the shear layer into surrounding fluid.

Like that in the round jet (see Chap. 2), this process looks similar to the development of Λ-structures in a boundary layer with their "heads" transported into the external flow (Reuter and Rempfer 2000). Though the roughness elements are located at one side of the nozzle only, the mushroom-like structures develop in both shear layers. However, the larger structures are observed in the layer modulated by

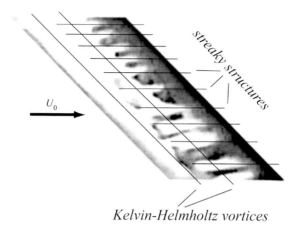

Fig. 3.2 Flow pattern in the shear layer of the free plane jet at $U_0 = 3$ m/s

Fig. 3.3 Smoke visualization of the jet cross sections at 5, 25, 45, 65 and 85 mm from the nozzle exit (*left* to *right*)

the roughness elements. The corresponding video (see multimedia file No. 3.6) also shows a vortex motion inside the mushroom-like structures which is not synchronized with the acoustic excitation and the periodicity of vortices. Flow visualization which was carried out in different cross and streamwise sections of the jet indicated that the Kelvin–Helmholtz vortices do not penetrate into this area, while mixing of the jet with ambient air is dominated by the behavior of these structures. Also note that the potential jet core is continuously diminished with the streamwise distance because of the growth of the shear layers thickness. At about sixfold increase of the jet Reynolds number from Re = 1300 to 8100 the mixing process appeared more effective, however, without qualitative variation of the observed instability phenomena (see Presentation Chap. 3: "Multimedia files Nos. 3.1–3.6") (http://extras.springer.com).

3.2   Three-Dimensional Perturbations
of the Wall Plane Jet

3.2.1   Methodology

Experimental data on the wall plane jet were obtained in the facility sketched in
Fig. 3.4. The jet originated at the nozzle exit 11×500 mm ($h \times l$) in size and
developed along a horizontal flat plate 2.1 m in length and 3.2 m in width.
A settling chamber equipped with a perforated plate, two honeycombs and a
deturbulizing grid was supplied with air by a fan. The turbulence level at the jet
origin measured in the band of 10 Hz–10 kHz at the Reynolds number
$\mathrm{Re} = U_0 h/v = 5000$, where U_0 is the jet core velocity and h is the width of nozzle
exit, was less than 0.05 %. Examination of the jet instability was performed through
flow visualization and hot-wire measurements similar to that described in Sect. 2.1.

3.2.2   Near Field of the Jet in Natural Experimental
Conditions

There are two characteristic layers distinguished in the jet, that is, the inner one
which is similar to a near-wall boundary layer and the external free shear layer
being far from the surface; in Fig. 3.4 they are marked "*1*" and "*2*", respectively.
The corresponding mean velocity profiles are shown in Fig. 3.5. Hot-wire measurements in the near-wall layer testify to a laminar flow with the mean velocity
distribution close to a Blasius profile (Fig. 3.5, *right*). Beginning approximately
from $x/h = 5$, where x is the streamwise distance measured from the nozzle exit
plane, high turbulence level in the upper part of the jet results in laminar-turbulent
transition in its internal region. In the external part of the jet (Fig. 3.5, *left*), the
mean velocity profiles normalized by the local maximum velocity and the
momentum thickness look similar in the range of $0.5 < x/h < 2.5$.

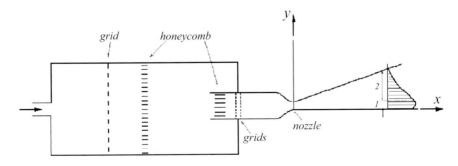

Fig. 3.4 Experimental facility for modeling of the wall plane jet

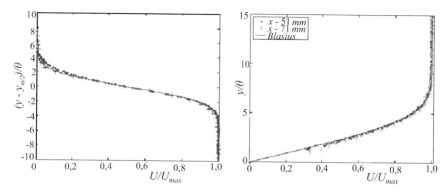

Fig. 3.5 Mean velocity distributions in the near field of the jet: the free shear layer (*left*) and the near-wall layer (*right*); $U_0 = 8$ m/s

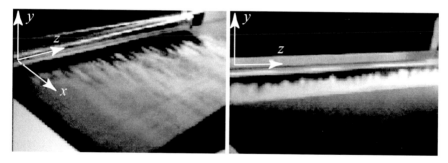

Fig. 3.6 Visualization of the wall plane jet in natural conditions: streamwise (*left*) and cross (*right*) sections (see Presentation Chap. 3: "Multimedia files No. 3.7 and 3.8") (http://extras.springer.com)

Qualitative data on the natural longitudinal structures generated at the nozzle exit were acquired by means of flow visualization so that the location of the perturbations and their scales were estimated. The corresponding flow patterns are shown in Fig. 3.6 (for more details see multimedia files No. 3.7 and 3.8). The three-dimensional disturbances are clearly seen together with the Kelvin–Helmholtz vortices in the x-z plane at the light sheet parallel to the wall (Fig. 3.6, *left*). Similar observations were done in a round jet (Monkewitz and Huerre 1982) where it was shown that the longitudinal perturbations exist between the primary vortices. Also, the disturbances are visible in the y-z plane at the light sheet perpendicular to the surface (Fig. 3.6, *right*). Most likely, their origination is due to perturbations arising in the settling chamber.

3.2.3 Controlled Transverse Modulation of the Jet

To stabilize the three-dimensional flow disturbances, five roughness elements each of 0.22-mm thickness, 15-mm length, and 7.5-mm width were pasted at the upper side of the inner nozzle surface near the exit. In this case, the forced longitudinal structures were generated at fixed positions. Then, the velocity perturbations were studied in detail through hot-wire measurements performed with a high spatio-temporal resolution. The results are presented in Fig. 3.7 (see also multimedia file No. 3.9); each flow pattern shown in the figure was reconstructed in a Matlab environment when processing the hot-wire data acquired in 3120 points. In Fig. 3.7a, the three-dimensional perturbations generated in the free shear layer and in Fig. 3.7b—those naturally appearing in the near-wall layer are shown as deviations of mean velocity from its value averaged over the transverse coordinate. As is found, the controlled disturbances of the free shear layer induce similar perturbations in the boundary layer, however, differing by their transverse scales.

To make clear the optimum scale of the perturbations and the effect of roughness size on their generation, the following tests were carried out. Four groups of the roughness elements, each of four identical ones, were examined at variation of the elements width as 5, 7.5, 10, and 12 mm. In each group, the distance between roughness elements was equal to their width, i.e., 5-mm elements were spaced at 5 mm, etc. The hot-wire data on the mean velocity perturbation produced by the above roughness configuration in the free shear layer of the jet are shown in Fig. 3.8. Here the elements width progressively increases from left to right. In the upstream sections, the most amplifying disturbances are found behind large roughness elements (10 and 12.5 mm) and have a preferable wave length in the transverse direction equal to 20–25 mm. Starting from about $x = 30$ mm the 2D Kelvin–Helmholtz vortices grow rapidly and the characteristic scale of the modulation becomes smaller.

At the jet core velocity increased up to $U_0 = 15$ m/s the laminar-turbulent transition was accelerated with some reduction of the transverse scale of

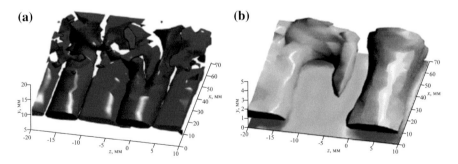

Fig. 3.7 Modulation of the free shear layer (**a**) and the boundary layer (**b**): mean velocity disturbance as isosurfaces corresponding to ±6 % of U_o; positive and negative values are in *dark blue* and *red* (*left*) and in *yellow* and *blue* (*right*), respectively

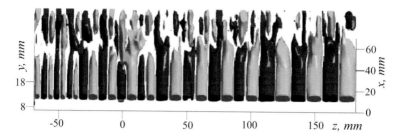

Fig. 3.8 Longitudinal disturbances excited in the free shear layer by the roughness elements at variation of their spacing: mean velocity disturbance as isosurfaces corresponding to ±5 % of U_0; positive and negative values are in *red* and *blue*, respectively; $U_0 = 8$ m/s

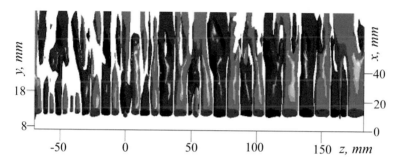

Fig. 3.9 Longitudinal disturbances excited in the free shear layer by the roughness elements at variation of their spacing: mean velocity disturbance as isosurfaces corresponding to ±5 % of U_0; positive and negative values are in *red* and *blue*, respectively; $U_0 = 15$ m/s

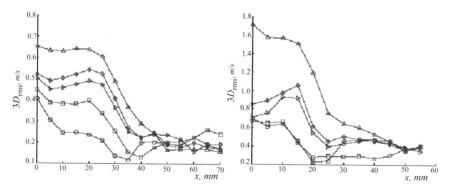

Fig. 3.10 Transverse scales of the longitudinal disturbances in the free shear layer at $U_0 = 8$ m/s (*left*) and $U_0 = 15$ m/s (*right*): 5 (O), 10 (□), 15 (∇), 20 (◊), and 25 mm (Δ)

perturbations, see Fig. 3.9. The latter is explained by diminution of the shear layer thickness with growth of the Reynolds number. One can observe the most amplifying scales determined through Fourier transform of the transverse mean velocity distributions in Fig. 3.10.

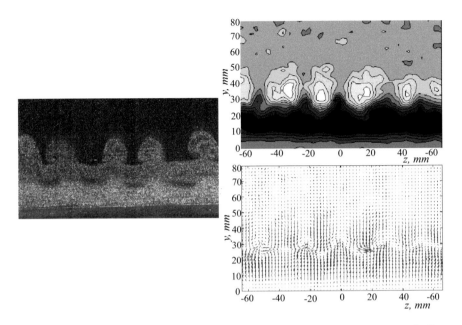

Fig. 3.11 Cross section of the jet at $x = 120$ mm: flow visualization (*left*), hot-wire data as isolines of the instantaneous streamwise velocity component (*top right*), and results of particle image velocimetry on instantaneous velocity vectors (*bottom right*)

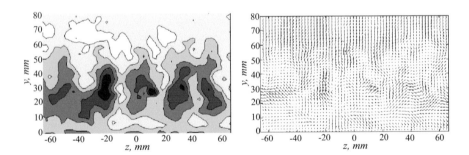

Fig. 3.12 Cross section of the jet at $x = 154$ mm: hot-wire data as isolines of the instantaneous streamwise velocity component (*left*) and results of particle image velocimetry on instantaneous velocity vectors (*right*)

The above data of flow visualization and hot-wire measurements are further supported by the results of particle image velocimetry (Litvinenko 2005) which provided velocity vectors in different sections of the jet. As an example, in Fig. 3.11 flow patterns obtained by different experimental methods are presented demonstrating a mushroom-like structure of velocity perturbations. Then, comparing Figs. 3.11 and 3.12 one can observe the breakdown of the quasi-periodic flow disturbances with increase of the streamwise coordinate.

Fig. 3.13 Velocity
isosurfaces of the
wall plane jet

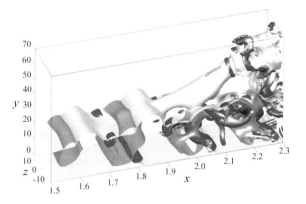

Further insight into dynamics of the wall plane jet was provided by Levin et al. (2005) in comparison to experimental data with results of the linear stability analysis predicting characteristics of the two-dimensional instability waves and of the stationary longitudinal perturbations subject to non-modal growth. Moreover, they studied the nonlinear stage of laminar-flow breakdown with direct numerical simulation making clear the interaction between two-dimensional instability waves and longitudinal (streaky) structures. As is seen in Fig. 3.13, the latter deform the primary vortices and are lifted up from the shear-layer region. More distinctly, the upward motion of the streaks resulting in a mushroom-shaped structure of velocity perturbations, similar to that shown in Fig. 3.11, is observed in Fig. 3.14 where several slices of the three-dimensional flow pattern of Fig. 3.13 are presented. Note that, first of all, turbulence of the wall jet is generated in its outer part and then spreads towards the wall. Obviously, the streaks play an important role in the jet evolution enhancing breakdown to turbulence.

3.2.4   *Acoustic Effect on the Jet Perturbations*

Controlled excitation of the Kelvin–Helmholtz instability in the free shear layer of the wall plane jet makes it possible to study influence of the frequency of two-dimensional oscillations on the generation and characteristics of the longitudinal structures. With this idea, the jet was subject to acoustic excitation at controlled frequencies by a loudspeaker placed above the jet at the nozzle exit plane. Thus it was found that frequency of the Kelvin–Helmholtz vortices have some effect on the transverse scale and intensity of the three-dimensional disturbances. Two most typical cases are compared in Fig. 3.15 (also we refer to multimedia files No's 3.10 and 3.11) presenting visualization data obtained with the light sheet. We observe that under the excitation at $f = 200$ Hz three-dimensional effects are less pronounced than at $f = 700$ Hz when three-dimensional flow structures at the transition to turbulence are well seen. The visualization results are in overall agreement with the hot-wire data shown in Fig. 3.16.

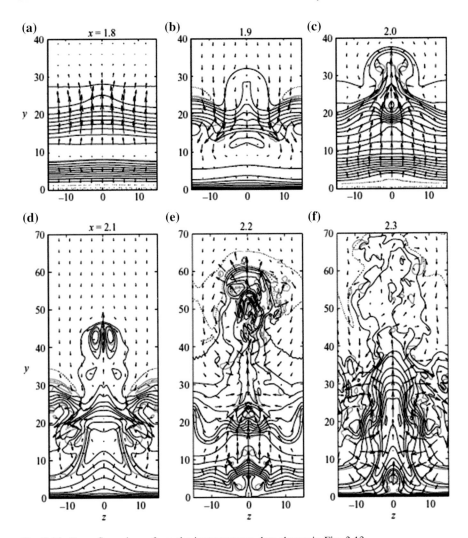

Fig. 3.14 Cross-flow planes from the instantaneous data shown in Fig. 3.13

Key points

- The dynamics of a near-wall plane jet is studied using both calculations and experiments. It is found that an arbitrary laminar wall jet can be successfully described by the solution of the boundary-layer equations and its solution is valid in the close downstream field of a nozzle.
- Afterwards, linear stability of the jet is investigated experimentally and theoretically. Comparison of the results of linear stability calculations with

Fig. 3.15 Laminar flow perturbations generated by roughness elements in the free shear layer of the jet forced by acoustic oscillations at 200 Hz (*left*) and 700 Hz (*right*) (see Presentation Chap. 3: "Multimedia files No. 3.10 and 3.11") (http://extras.springer.com)

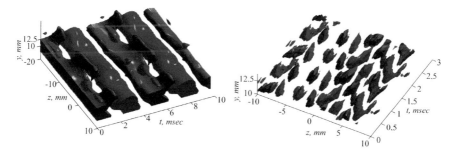

Fig. 3.16 Velocity perturbations of the free shear layer of the jet disturbed by the longitudinal structures under acoustic forcing: $f = 200$ Hz, $x = 41$ mm, isosurface levels are 0.3 U_0 (*left*); $f = 700$ Hz, $x = 21$ mm, isosurface levels are 0.4 U_0 (*right*); positive and negative values are in *red* and *blue*, respectively; $U_0 = 8$ m/s

experiments shows that the theory is able to predict the most amplified frequency of the periodical waves and the most amplified scale of the streaks.

- These spatio-temporal scales are dominating in experiments with unforced jet. Furthermore, linear stability theory demonstrates rather high instability of the flow to non-modal streaks and it seems that this mechanism is responsible for generation of initial of three-dimensionality of the jet breakdown.

- Additional support of this conclusion is the excellent agreement between the calculated and measured amplitude functions of the streak. The calculations indicate that the optimal disturbance represents streamwise vortices, which cause the formation of streaks by the so-called lift-up effect.

- Finally, experimental studies of the nonlinear stage of the laminar flow breakdown are supported by direct numerical simulation (DNS). Three-dimensional simulations with random and coherent forcing are provided and both simulations clearly demonstrated that growing streaks are important for the breakdown process. As experiments and simulations show, very strong amplification of streaks occurs at the stage of non-linear interaction between streaks and two-dimensional waves.

References

Levin O, Chernoray VG, Lofdahl L, Henningson DS (2005) A study of the Blasius wall jet. J Fluid Mech 539:313–347

Litvinenko MV (2005) Formation and role of longitudinal structures at laminar-turbulent transition in jets. Ph.D. thesis, Novosibirsk State Technical University, in Russian

Monkewitz PA, Huerre P (1982) Influence of the velocity ratio on the spatial instability of mixing layers. Phys Fluids 25(7):1137–1143

Reuter J, Rempfer D (2000) A hybrid spectral/finite-difference scheme for the simulation of pipe-flow transition. In: H Fasel, WS Saric (eds) Laminar-turbulent transition. Springer, Berlin, pp 383–390

Westin KJA, Boiko AV, Klingmann BGB, Kozlov VV, Alfredsson PH (1994) Experiments in a boundary layer subjected to free stream turbulence. Part 1. Boundary layer structure and receptivity. J Fluid Mech 281:193–218

Chapter 4
Round Jet Instability Affected by Initial Conditions

Obviously, evolution of a jet depends on the initial conditions at the nozzle exit. In particular, varying the nozzle geometry and, hence, the velocity distribution near its exit, it is possible to modify instability of the jet and its dynamics. Such an approach to control of jet flows is the main subject of this chapter; see also the supplementing material Chap. 4: "Multimedia files Nos. 4.1–4.3" (http://extras.springer.com).

4.1 Experimental Technique

In the present case, the test facility illustrated in Figs. 4.1 and 4.2 is distinguished by extension pipes of different length (430, 870, 1300, and 4000 mm) which could be attached to a nozzle of 20-mm diameter (d). In this way, it was possible to change the initial mean velocity profile assuming that, at a sufficiently long pipe, a velocity distribution close to the Hagen–Poiseuille profile could be obtained.

Evolution of both laminar and turbulent jets was examined at one and the same jet core velocity $U_0 = 5$ m/s corresponding to the Reynolds number Re $= U_0 d / v \approx 6700$. The turbulent state was generated by a sand paper strip of 5-mm width glued onto the inner outlet surface of the nozzle. The effect of external acoustic forcing on the jet was also investigated. Flow visualization and hot-wire measurements were carried out similarly to that described in the previous sections.

4.2 Instability of the Laminar Jet

The above experimental facility made it possible to examine the jet evolution depending on the initial velocity distribution that is discussed below in more detail.

© The Author(s) 2016
V.V. Kozlov et al., *Visualization of Conventional and Combusting Subsonic Jet Instabilities*, SpringerBriefs in Applied Sciences and Technology, DOI 10.1007/978-3-319-26958-0_4

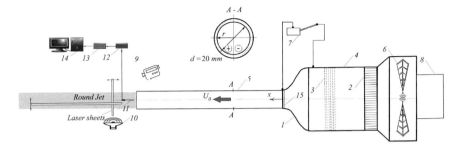

Fig. 4.1 Experimental set-up: Vitoshinsky nozzle (*1*), honeycombs (*2*), grids (*3*), settling chamber (*4*), extension pipe (430, 870, 1300, or 4000-mm length) (*5*), fan (*6*), liquid manometer (*7*), smoke generator (*8*), digital camera (*9*), dynamic loudspeaker (*10*), hot-wire probe (*11*), hot-wire anemometer (*12*), ADC (*13*), PC (*14*), roughness strip (*15*)

Fig. 4.2 General view of the round jet facility with the 4000-mm extension pipe (**a**) and the original (short) nozzle (**b**)

4.2.1   Basic Configuration of the Nozzle

Cross sections of the mean velocity (U) and fluctuations (u') of the jet issued from the original nozzle (without an extension pipe) along with visualization results are shown in Fig. 4.3. It is seen in Fig. 4.3*a* that the distribution of the mean flow velocity across the jet is a top-hat one. A minor dip in the core region is, most probably, due to the ring vortices rolling up in the shear layer. The maximum level of velocity perturbations is found at the jet periphery, see Fig. 4.3*b*. Near the nozzle it makes 1 % of U_0 and then gradually increases downstream up to about 4.5 % of U_0. In the jet core, the disturbances are much lower with the amplitude varying in the measurement domain from 0.2 to 0.6 % of U_0. The above velocity distributions are typical of classical round jets. Considering the flow visualization data, one can distinctly observe the ring vortices (see Fig. 4.3*c*) and three-dimensional bursts periodically spaced over the azimuthal coordinate (see Fig. 4.3*d*) generated at the interaction between the vortices and the longitudinal perturbations as discussed in Chap. 2.

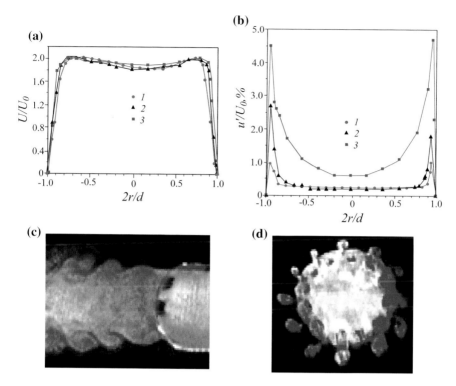

Fig. 4.3 Distributions of mean velocity (**a**) and perturbations (**b**) in the jet cross sections at $x = 2$ (*1*), 10 (*2*), and 20 mm (*3*) from the nozzle exit; flow visualization of the streamwise (**c**) and cross (**d**) sections (see Presentation Chap. 4: "Multimedia file No. 4.1") (http://extras.springer.com)

4.2.2   The Nozzle Modified by the Extension Pipes

Consider now what happens with the jet flow at modification of the nozzle by the 430-mm long extension pipe. Figure 4.4*a* shows that the mean-velocity profile tends to a parabolic shape, although a plateau around the jet axis is still observed. At the same time, the velocity disturbances go down in the shear layer and in the jet core, compare Fig. 4.4*b* to Fig. 4.3*b*. As a result, a pronounced laminar flow region starts growing behind the nozzle exit that is illustrated in Fig. 4.4*c*.

Proceeding with the extension pipe of the length equal to 870 mm, we observe further transformation of the mean velocity profiles towards the parabolic shape (Fig. 4.5*a*) and reduction of the perturbation amplitude (Fig. 4.5*b*). The laminar portion of the jet becomes more extended so that the transition to turbulence is obviously delayed (Fig. 4.5*c*). Finally, at the exit of the 4000-mm long pipe almost a parabolic velocity profile is generated, see Fig. 4.6. In this case the laminar flow region is the most prominent, spreading up to about 10*d* behind the pipe cut.

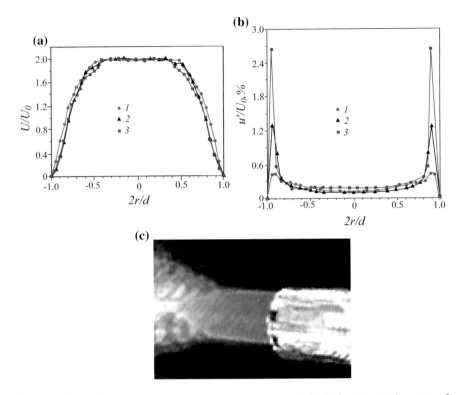

Fig. 4.4 Distributions of mean velocity (**a**) and perturbations (**b**) in the jet cross sections at $x = 2$ (*1*), 10 (*2*), and 20 mm (*3*) behind the exit of the 430-mm extension pipe; flow visualization (**c**) (see Presentation Chap. 4: "Multimedia file No. 4.2") (http://extras.springer.com)

Thus, the experiments show a possibility to control the flow structure of a round jet modifying the initial conditions at the nozzle exit through lengthening of nozzle channel, i.e. the transformation of the mean velocity distribution from the classical top-hat one to its parabolic profile.

4.2.3   Acoustic Influence upon the Modified Jet Flow

It was already noted in Sect. 2.5 that external acoustic forcing of the round jet affects the generation of the ring vortices and their interaction with the three-dimensional disturbances. A somewhat similar situation is observed when the 430-mm long extension pipe is attached to the basic nozzle. As is seen in Fig. 4.7, well behind the pipe outlet, the vortex structures become visible with their scale changing about two times depending on the excitation frequency.

In the case of the 4000-mm long pipe, acoustic forcing appeared inefficient when applied to the laminar part of the jet, at least, at the sound frequencies up to several

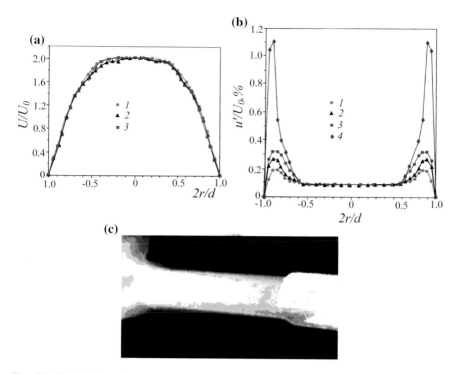

Fig. 4.5 Distributions of mean velocity (**a**) and perturbations (**b**) in the jet cross sections at $x = 2$ (*1*), 10 (*2*), 20 (*3*), and 60 mm (*4*) behind the exit of the 870-mm extension pipe; flow visualization (**c**) (see Presentation Chap. 4: "Multimedia file No. 4.3") (http://extras.springer.com)

kilohertz and the pressure level up to 100 dB. At the same time, acoustic excitation was still useful at modification of vortex structures emerging far downstream of the pipe cut as illustrated in Fig. 4.8 and look like helical vortices. One expects prevalence of such perturbations in the flow with the present mean-velocity profile shown in Fig. 4.6 instead of the axisymmetric Kelvin–Helmholtz vortices typical of the jet with a thin shear layer issuing from the short nozzle (Cohen and Wignanski 1987).

4.3 Modeling of the Turbulent Jet

The roughness strip at the outlet of the basic nozzle resulted in a considerable variation of the mean-velocity and disturbance characteristics of the jet as compared to those of the laminar state, see Fig. 4.9. At first, we turn to the mean-flow data of Fig. 4.9*a* presenting the velocity distributions measured near the exit of the extension pipes of lengths equal to 870, 1300, and 4000 mm; in this plot, the profiles are normalized by the mean velocity of the jet U_0. Behind the pipes of

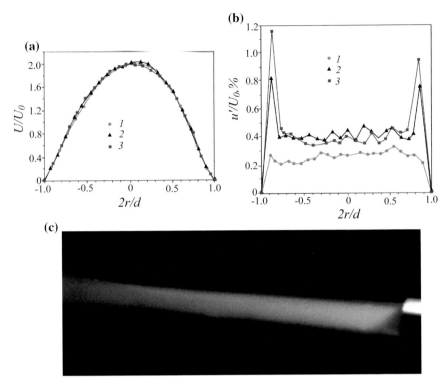

Fig. 4.6 Distributions of mean velocity (**a**) and perturbations (**b**) in the jet cross sections at $x = 2$ (*1*), 10 (*2*), and 20 mm (*3*) behind the exit of the 4000-mm extension pipe; flow visualization (**c**)

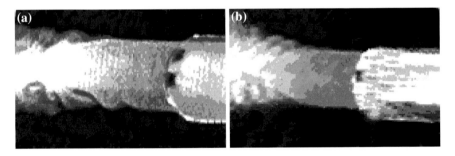

Fig. 4.7 Visualization of the laminar jet emanating from the 430-mm extension pipe under acoustic excitation at 100 (**a**) and 250 Hz (**b**)

the length equal to 1300 and 4000 mm, the turbulent mean-flow profile is found. In these flow configurations, the boundary layer perturbed by the roughness strip grows on the pipe wall over a large distance, thus spanning the entire cross section of the exit. Otherwise, at the pipe length reduced to 870 mm, the jet is not fully

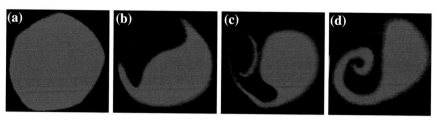

Fig. 4.8 Visualization of the jet cross sections at $x = 210$ (**a**), 240 (**b**), 300 (**c**), and 350 mm (**d**) behind the exit of the 4000-mm extension pipe

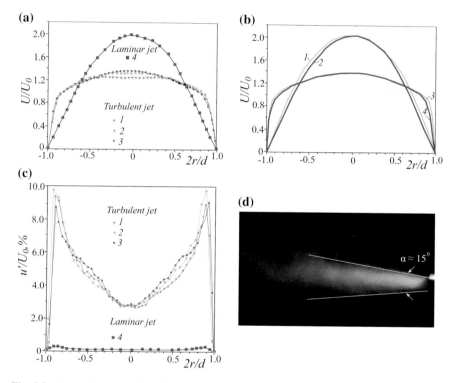

Fig. 4.9 Mean-velocity profiles of the jet perturbed by the roughness strip at the extension pipes of 870 (*1*), 1300 (*2*), and 4000 mm (*3*) and of the laminar jet emanating from the 4000-mm pipe (*4*), all measured at 2 mm behind the pipe cut (**a**); mean-velocity distributions of the laminar and turbulent jets (*2* and *4*) as compared to those of the pipe flow (*1* and *3*, it is taken from the work of Durst and Bulent 2006) (**b**); velocity perturbations in the jet cross section, for notation see **a** (**c**); visualization of the turbulent jet behind the 4000-mm pipe (**d**)

turbulent with a plateau of the mean velocity in the core region. For comparison, in Fig. 4.9*a* the laminar-flow profile behind the 4000-mm extension pipe is shown. Then, in Fig. 4.9*b* the present mean-velocity distributions of the laminar and turbulent jets are compared to similar results for a pipe flow (Durst and Bulent 2006) which are in overall agreement.

Radial distributions of velocity perturbations in the jet at different length of the extension pipe are demonstrated in Fig. 4.9c similarly to the mean-flow data of Fig. 4.9a. In the case of the roughness strip, the amplitude of disturbances becomes an order of magnitude larger than that in the laminar flow both at the jet periphery and in the core region. Note that the disturbance profiles of the turbulent jet (marked 2 and 3 in Fig. 4.9c) and of that approaching the turbulent mode (marked 1) are practically coincide.

Comparing visualization results shown in Figs. 4.6c and 4.9d, one can observe much different flow patterns of the laminar and turbulent jets. While the first of them keeps almost constant width over a large streamwise distance, the second one spreads at an angle of about 15 degrees. However, the present experiments indicated that the basic features of the jet dynamics discussed in Chap. 2 including generation of the ring vortices and their interaction with the longitudinal perturbations resulting in the three-dimensional bursts periodically spaced along the azimuthal coordinate apply to the jet perturbed by the roughness strip, as well.

Key points

- The flow structure and evolution of a round jet are strongly affected by the initial conditions at the jet origin.
- Through a proper modification of the nozzle profile (hence, the radial mean-velocity distribution) the laminar part of the jet becomes much more extended.
- In the laminar jet the acoustic effect on the flow pattern becomes visible well behind the nozzle outlet where the ring vortices start to grow.
- Profiling the nozzle combined with forced turbulization of its boundary layer makes it possible to create a fully turbulent flow immediately at the jet origin.

References

Cohen J, Wignanski I (1987) The evolution of instabilities in axisymmetric jet. Pt. 1. The linear growth of disturbances near the nozzle. J Fluid Mech 176:191–219
Durst F, Bulent U (2006) Forced laminar-turbulent transition of pipe flows. J Fluid Mech 560:449–464

Chapter 5
Origination and Evolution of Coherent Structures in Laminar and Turbulent Round Jets

In this chapter, dynamics of laminar and turbulent round jets is under further consideration. Experimental data testify to similarities of the generation and development of coherent structures in both cases. Also, almost one and the same response of the jets to external acoustic oscillations is demonstrated. Supplementary material to this chapter is given by the multimedia files titled in Chap. 5 (http://extras.springer.com).

5.1 Experimental Set-up

The sketch of the test facility and its general view are presented in Figs. 5.1 and 5.2. Basically, it is designed as a Vitoshinsky nozzle with the outlet diameter $d = 20$ mm (*1*) equipped with a honeycomb (*2*) and a set of grids (*3*) in a settling chamber (*4*). The jet flow was generated at the cut of an extension pipe of the length equal to 150 mm (*5*) attached to the nozzle. Using a sand paper strip of 5-mm width (*6*) at the nozzle exit, a perturbed jet with the mean velocity profile approaching the turbulent one was created, see Sect. 4.3. To induce streamwise elongated disturbances in the jet shear layer, eight roughness elements each of 4-mm width, 14-mm length, and 1-mm thickness were periodically spaced at the exit of the extension pipe. The passage frequency of the ring vortices was stabilized using external acoustic oscillations generated by a dynamic loudspeaker (*10*). Like that in the previous chapters, flow visualization and hot-wire measurements were carried out. During visualization, smoke was injected from the fan side (*8*) and the flow patterns illuminated by a laser sheet were recorded by a digital camera (*9*). Velocity characteristics were determined using a constant-temperature anemometer and one-wire probes with computer data processing, (*11*)–(*14*) in Fig. 5.1.

The laminar and turbulent jets were examined at a constant velocity at the center of the pipe exit $U_0 = 4$ m/s which was controlled with a pitot-static tube and a liquid manometer (*7*). The corresponding Reynolds number was Re $= U_0 d/\nu \approx 5300$.

© The Author(s) 2016
V.V. Kozlov et al., *Visualization of Conventional and Combusting
Subsonic Jet Instabilities*, SpringerBriefs in Applied Sciences and Technology,
DOI 10.1007/978-3-319-26958-0_5

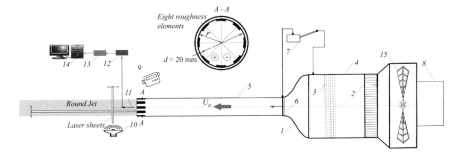

Fig. 5.1 Experimental set-up: Vitoshinsky nozzle (*1*), honeycomb (*2*), grids (*3*), settling chamber (*4*), extension pipe (*5*), sand paper strip (*6*), liquid manometer (*7*), smoke generator (*8*), digital camera (*9*), dynamic loudspeaker (*10*), hot-wire probe (*11*), hot-wire anemometer (*12*), ADC (*13*), PC (*14*), fan (*15*)

Fig. 5.2 General view of the jet facility

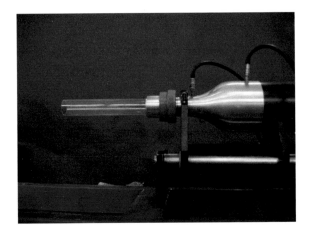

5.2 Velocity Characteristics

Quantitatively, the laminar and turbulent jets are compared in Figs. 5.3 and 5.4 showing the profiles of mean velocity and non-stationary perturbations measured in several cross sections. The difference between two flow regimes is demonstrated by the mean-flow data and, more apparently, by the disturbance distributions. In the turbulent jet, high amplitude of the perturbations is found in the shear layer immediately behind the jet origin (Figs. 5.4*a*–*d*), otherwise, the same level of disturbances is observed further downstream when the laminar jet undergoes the transition to turbulence (Fig. 5.4*e*).

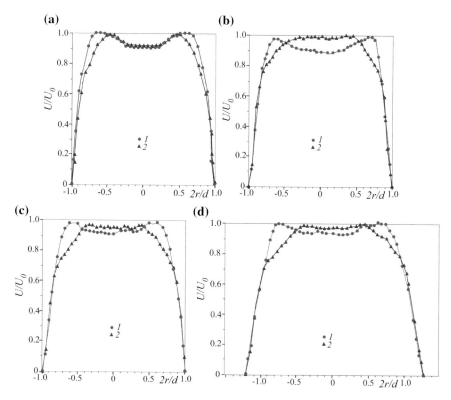

Fig. 5.3 Mean velocity distributions of the laminar (*1*) and turbulent (*2*) jets measured at $x = 1$ (**a**), 2 (**b**), 5 (**c**), and 20 mm (**d**) behind the exit of the extension pipe

5.3   Visualization of the Jet Perturbations

Flow patterns of the laminar jet are illustrated in Fig. 5.5. In the general view (Fig. 5.5*a*) one can observe origination of the ring vortices and the laminar flow breakdown further downstream. Then, longitudinal structures of perturbations induced at the jet periphery by the roughness elements are shown in Fig. 5.5*b*. Interaction of the disturbances resulting in the generation of three-dimensional bursts periodically spaced in the cross section and the jet turbulization are clearly seen in Fig. 5.5*c*.

To a certain extent, one expects similar dynamics of the turbulent jet (see Fig. 5.6) that is supported by the visualization results given in Fig. 5.7. In this case, the turbulent flow is found close to the outlet of the extension pipe (Fig. 5.7*a*). The streamwise elongated disturbances generated by the roughness elements are visible in Fig. 5.7*b*. The flow structure in the cross sections (Fig. 5.7*c*) is not so pronounced as compared to the laminar jet (Fig. 5.5*c*), however, the ring vortices and the

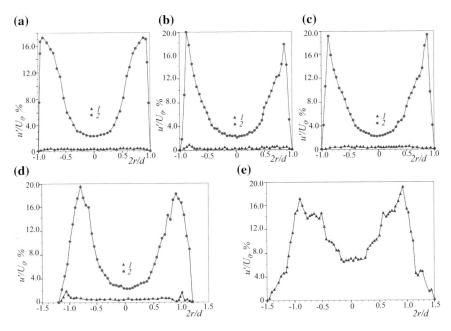

Fig. 5.4 Rms amplitudes of velocity perturbations of the laminar (*1*) and turbulent (*2*) jets measured in the frequency band of 30–80 Hz at $x = 1$ (**a**), 2 (**b**), 5 (**c**), 20 (**d**) and 60 mm (**e**) behind the exit of the extension pipe

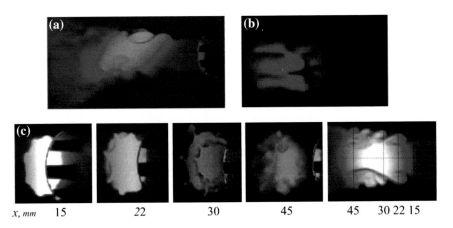

x, mm   15   22   30   45   45   30 22 15

Fig. 5.5 Visualization of the laminar round jet: general view (**a**); longitudinal disturbances (**b**); cross sections at different distances from the extension pipe outlet (**c**)

longitudinal perturbations are still found there. More distinctly the ring vortices are observed in Fig. 5.7*d* presenting an enlarged flow pattern near the jet origin.

The above results reveal that the evolution of both laminar and turbulent jets is dominated by the generation of the ring vortices and the longitudinal structures

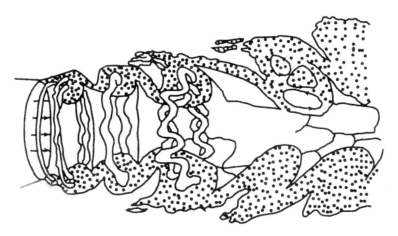

Fig. 5.6 Origination of the coherent structures in a round turbulent jet (Guinevskiy et al. 2001)

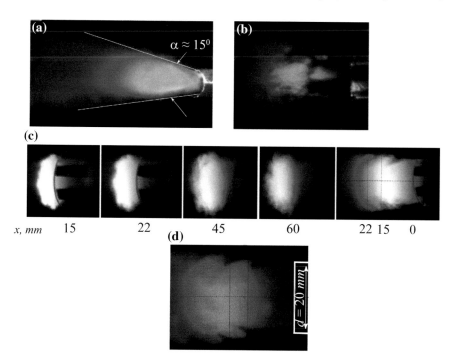

Fig. 5.7 Visualization of the turbulent round jet: general view (**a**); longitudinal disturbances (**b**); cross sections at different distances from the extension pipe outlet (**c**); upstream part of the jet (**d**)

interacting with each other. Primarily, this conclusion became available due to the low Reynolds number of the present experiments which provided high spatial resolution of the disturbed flow pattern and observation of the controlled velocity perturbations.

5.4 Response of the Jets to Acoustic Oscillations

To compare the receptivity of the laminar and turbulent jets to external acoustic excitation, they were forced by weak harmonic oscillations generated by a loud-speaker as is sketched in Fig. 5.1. The loudspeaker was located near the exit of the extension pipe and radiated sound waves at the frequency of 180 Hz transversally to the flow direction. Then, amplitude of the ring vortices originating in the jets at the forcing frequency was determined through ensemble averaging of the hot-wire traces.

Maximum amplitudes of the vortical perturbations of the laminar and turbulent jets excited at a fixed sound pressure level are presented in several cross sections in Fig. 5.8. Here the data acquired in the jet shear layer are shown; the disturbances of

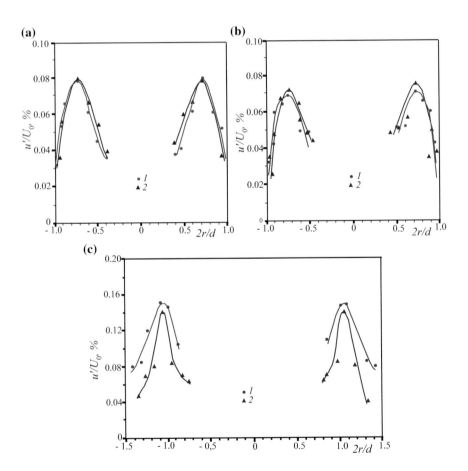

Fig. 5.8 Amplitude distributions of the excited flow perturbations in cross sections of the turbulent (*1*) and laminar (*2*) jets at $x = 2$ (**a**), 5 (**b**) and 20 mm (**c**) behind the exit of the extension pipe

the core region are much lower and not included. In both cases, almost one and the same response of the jets to periodic excitation is observed. Also, one expects linearity of the generated disturbances at their amplitude in the present experimental conditions as small as about 0.1 % of U_0. The above data confirm once again that, in different states of the basic flow, the main features of the origination and behavior of the dominant perturbations in round jets are the same.

Key points

- Main features of the laminar jet instability including generation of the ring vortices and their interaction with the longitudinal structures of perturbations apply to the turbulent jet, as well.
- Quantitatively, almost the same response of laminar and turbulent jets to external acoustic forcing at generation of the ring vortices is found.

Reference

Guinevskiy AS, Vlasov EV, Karavosov RK (2001) Acoustic control of turbulent jets. Nauka, Moscow, in Russian

Chapter 6
Plane Jets Affected by Initial Conditions and Acoustic Perturbations

In this chapter, results of experimental studies on the influence of initial conditions at the nozzle exit and acoustic perturbations on a plane jet structure and
characteristics of its evolution and stability are considered. Basic features of the
laminar and turbulent plane jets at one and the same Reynolds number are
demonstrated. It is found that the jets are subjected to sinusoidal oscillations
suppressing the varicose mode of instability. Also, the focus is on interaction of
the longitudinal structures of flow perturbations generated on one side of the
nozzle with large-scale two-dimensional vortices of the laminar plane jet
resulting in origination of Λ or Ω-shaped vortex structures. For supplementary
material to this chapter see Chap. 6: "Multimedia files Nos. 6.1–6.9" (http://
extras.springer.com).

6.1 Experimental Set-up

The plane-jet facility used for the observations discussed in what follows is shown
in Figs. 6.1 and 6.2. In general, this set-up is similar to that of the experiments on
round jets discussed in the previous chapter. In the present case, the cross sections
of the settling chamber and the nozzle outlet are 0.4–0.4 and 0.014–0.4 m making a
contraction ratio of about 29. To obtain uniform velocity distribution before the
contraction, the settling chamber was additionally equipped with a flow-dividing
plate and a perforated one, see Fig. 6.1.

The initial velocity profile was modified by an extension channel of a length
equal to 3540 mm attached to the basic nozzle. The turbulent flow was generated
using a sand paper strip at the outlet of the basic nozzle and the longitudinal
structures of flow perturbations were induced by roughness elements spaced at the

© The Author(s) 2016
V.V. Kozlov et al., *Visualization of Conventional and Combusting
Subsonic Jet Instabilities*, SpringerBriefs in Applied Sciences and Technology,
DOI 10.1007/978-3-319-26958-0_6

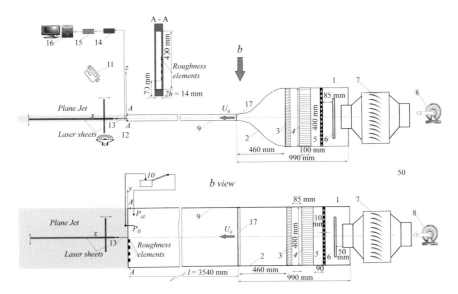

Fig. 6.1 Experimental set-up: settling chamber (*1*); Vitoshinsky nozzle (*2*); grids (*3*); honeycomb (*4*); perforated plate (*5*); dividing plate (*6*); fan (*7*); smoke generator (*8*); extension channel (*9*); liquid manometer (*10*); digital camera (*11*); dynamic loudspeaker (*12*); hot-wire probe (*13*); hot-wire anemometer (*14*); ADC (*15*); PC (*16*); sand paper strip (*17*)

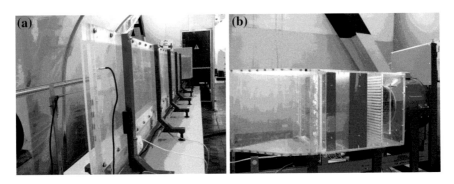

Fig. 6.2 Plane-jet facility with the extension channel (**a**), the basic nozzle with the settling chamber and the fan (**b**)

exit of the extension channel. The laminar and turbulent jets were examined through flow visualization and hot-wire measurements at one and the same jet core velocity $U_0 = 3.7$ m/s, that is, Re $= U_0 h/\nu \approx 1700$ where $h = 0.007$ m is the half-width of the nozzle exit.

6.2 Instability of the Laminar Plane Jet with a Parabolic Mean Velocity Profile

Obviously, the mean velocity profile in the laminar region of a plane jet is affected by the conditions at the nozzle outlet. As is known, the mean flow pattern at the exit of a plane channel depends on the parameter vl/h^2U_0, where l is the channel length, h is its half-width, U_0 is flow velocity at the channel center and v is kinematic viscosity (Schlichting 1934). Thus, the initial velocity distribution of the laminar jet is close to a parabolic one provided that the extension channel is long enough; in the present set-up this parameter was about 0.27. Otherwise, thin boundary layers on the channel walls result in strong velocity gradients at the jet periphery and almost constant velocity in the jet core. Such mean velocity distribution is often known as a "top-hat" profile. The top-hat and parabolic mean velocity profiles of the present experiments measured behind the basic nozzle and the extension channel are shown in Fig. 6.3.

Amplifying oscillations of laminar plane jets are associated with symmetric and anti-symmetric instability modes (Sato 1960). Being found in different configurations of shear layers, similar perturbations are also referred to as varicose and sinusoidal modes, respectively (Boiko et al. 2002, 2012). In the case of top-hat velocity profile of the jet, evolution of flow disturbances appears as a competition of these two instabilities. The symmetric mode prevails at the initial stage of the laminar jet breakdown. Further downstream, the mean velocity profile tends to a parabolic one so that the anti-symmetric oscillations start to grow. In the present conditions, the parabolic mean velocity profile is generated directly at the jet origin; therefore the anti-symmetric mode seems to dominate the flow pattern.

This expectation is confirmed by the visualization data of Fig. 6.4 obtained in the central section of the laminar plane jet issuing from the extension channel in natural conditions and under controlled external acoustic excitation at a constant sound pressure level. Quantitatively, the disturbances visualized in Fig. 6.4 are

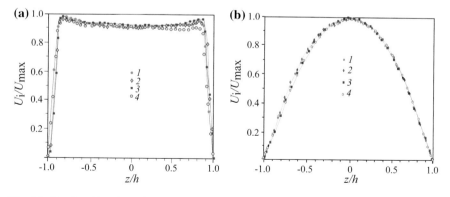

Fig. 6.3 Top-hat (**a**) and parabolic (**b**) mean velocity profiles of the laminar plane jet at $x = 2$ (*1*), 5 (*2*), 9 (*3*), and 15 mm (*4*) behind the basic nozzle/extension channel outlet

Fig. 6.4 Visualization of the laminar plane jet in natural conditions (*1*) and under external acoustic forcing at frequencies f = 30 (*2*), 40 (*3*), 50 (*4*), 60 (*5*), 70 (*6*), 90 (*7*) and 150 Hz (*8*) (see Presentation Chap. 6: "Multimedia files No. 6.1 and 6.2") (http://extras.springer.com)

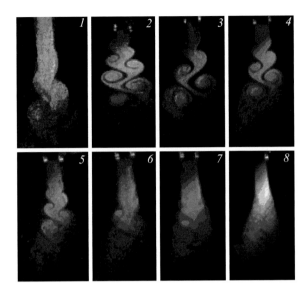

Table 6.1 Characteristics of the laminar jet oscillations

Conditions	Frequency (f, Hz)	Wave length (λ, mm)	Propagation velocity ($c = f\lambda$, m/s)	Non-dimensional values	
				Sh = $2fh/U_0$	c/U_0
Natural, Fig. 6.4 (*1*)	30	50	–	0.12	–
Forced, Fig. 6.4 (*2–8*)	30	50	1.5	0.12	0.40
	40	40	1.6	0.16	0.43
	50	35	1.75	0.19	0.47
	60	29	1.7	0.23	0.46
	70	25	1.75	0.27	0.47
	90	18	1.62	0.35	0.44
	150	11	1.65	0.57	0.45

compared in Table 6.1. In natural conditions, the perturbation frequency is about Sh = $2f\,h/U_0$ = 0.12, which is in good agreement with calculation and experimental results by Yu and Monkewitz (1990, 1993). Hussain and Thompson (1980) argued that, in the range of low Strouhal numbers, the plane jet appears as a non-dispersive waveguide. In the present case, this applies to Sh = 0.12 − 0.19, that is, to the natural perturbations and those induced by acoustic forcing at frequencies up to 50 Hz.

Propagation velocities of the perturbations indicated in Table 6.1 are shown in Fig. 6.5 reproducing a diagram plotted by Sato (1960) with his experimental data

Fig. 6.5 Propagation velocities of plane-jet oscillations: theory (*lines*) and experiment (*circles*) (Sato 1960), the present results (*triangles*)

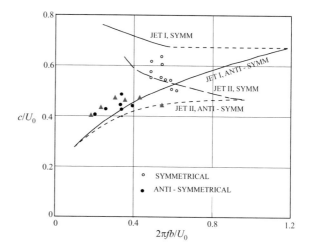

and theoretical curves for symmetric and anti-symmetric disturbances at different mean velocity profiles. Oscillation frequency is normalized here using half-breadth of the jet b and flow velocity at the center plane U_0. An overall agreement between the present experimental results and calculations for a plane jet with a parabolic mean velocity profile (JET II) is found.

As found in Fig. 6.4, quite different acoustic effects on the jet oscillations can be observed. The low-frequency forcing (images *2–5*) stabilizes the anti-symmetric disturbances and makes them more pronounced as compared to that in natural conditions. At increase of the excitation frequency, the oscillations become visible at a larger distance from the cut of the extension channel and turn to the symmetric mode (image *7*). The last observation is in agreement with the conclusion by Sato (1960) that the high-frequency perturbations generated in a plane jet by sound are symmetric while the low-frequency oscillations are anti-symmetric.

The results shown above were obtained in the transverse acoustic field. Additionally, to excite the low-frequency symmetric disturbances of the jet, sound waves radiated in the flow direction by a loudspeaker placed in the settling chamber were tried. However, in this case a symmetric flow pattern was observed only close to the exit of the extension channel over short time intervals, while still prevailing were the anti-symmetric oscillations, see Fig. 6.6.

Coming back to Fig. 6.4 we also notice a profound effect of acoustic excitation on the overall flow configuration, that is acceleration of transverse spreading of the jet and even its splitting at high-frequency forcing (see images *7* and *8*).

Fig. 6.6 Symmetric (**a**) and
anti-symmetric
(**b**) perturbations of the
laminar plane jet affected by
the acoustic waves radiated
from the settling chamber

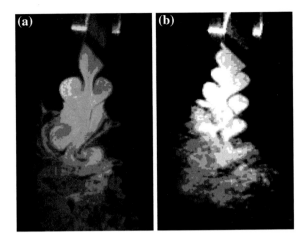

6.3   Longitudinal Structures of Perturbations at the Breakdown of a Plane Jet

A significant contribution of the longitudinal structures of disturbances to laminar jets breakdown has already been emphasized in the previous chapters. The same applies to the plane jet with a top-hat mean velocity profile at the nozzle exit. As an illustration, Fig. 6.7 shows the origination of the Λ- or Ω-like vortices in the shear layers of the jet due to natural stationary longitudinal perturbations (Kozlov et al. 2002).

In such case, thin shear layers on both sides of the jet as well as their disturbances evolve independently of each other. Thus, excitation of the longitudinal

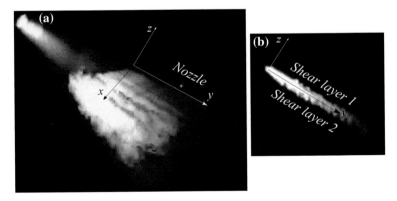

Fig. 6.7 Breakdown of the laminar plane jet with a top-hat mean velocity profile at the nozzle exit: general view (**a**), Λ- or Ω-like vortices visualized in a cross section (**b**) (see Presentation Chap. 6: "Multimedia file No. 6.3") (http://extras.springer.com)

structures by controlled roughness elements located at one side of the nozzle results in the Λ- or Ω- like vortices in the corresponding shear layer only, while in the other one the naturally occurring structures are much weaker (Boiko et al. 2005). This is well seen in Fig. 6.8 where the flow region perturbed by the roughness elements is marked as "*Shear layer 2*".

When turning to the laminar plane jet with a parabolic mean velocity profile, two aspects are noticeable. One of them is non-zero shear all across the jet and the other is the sinusoidal instability which initiates the anti-symmetric vortex street dominating the flow pattern sketched in Fig. 6.9. In such conditions, interaction of the longitudinal structures naturally emerging at the nozzle exit with the cylindrical vortices under acoustic excitation at $f = 30$ Hz is demonstrated in Fig. 6.10. A result of the interaction is the generation of the Λ- or Ω-like vortices which are shown by images *1–4* taken in the jet cross sections while moving downstream from the nozzle. The breakdown of vortices with opposite rotation is presented in parts *a* and *b*, respectively.

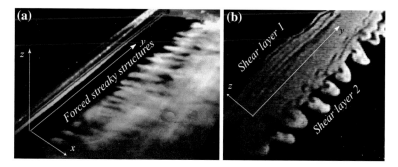

Fig. 6.8 Laminar plane jet with a top-hat mean velocity profile modified by controlled longitudinal structures: general view (**a**), Λ- or Ω-like vortices generated in the shear layer disturbed by the roughness elements (**b**) (see Presentation Chap. 6: "Multimedia files Nos. 6.4–6.7") (http://extras.springer.com)

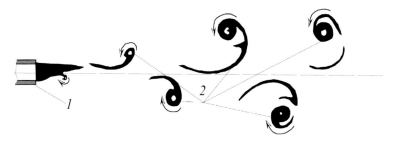

Fig. 6.9 Vortex street due to the sinusoidal instability of the laminar plane jet with a parabolic mean velocity profile at the exit of the extension channel: outlet of the channel (*1*) and counter rotating vortices (*2*)

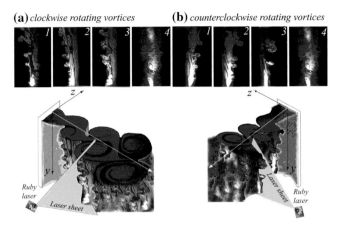

Fig. 6.10 Interaction of the natural longitudinal structures of disturbances with the clockwise (**a**) and counterclockwise (**b**) rotating vortices in the laminar plane jet with the parabolic mean velocity profile; in the bottom is the scheme of visualization (see Presentation Chap. 6: "Multimedia file No. 6.8") (http://extras.springer.com)

Further, dynamics of the jet in the presence of an isolated controlled roughness element at the outlet of the extension channel is illustrated in Fig. 6.11. As is sketched in Fig. 6.11*b*, a pair of stationary longitudinal structures is generated at the element. Interacting with the vortex street, they are curved and induce the Λ -or Ω-like

Fig. 6.11 Interaction of the forced longitudinal perturbation with the cylindrical vortices: two neighboring clockwise (*1, 2*) and counterclockwise (*3, 4*) rotating vortices (**a**); sketch of the vortex street with the exit of the extension channel (*1*), roughness element (*2*), longitudinal structures (*3*), and 3D bursts of the cylindrical vortices (*4, 5*) (**b**)

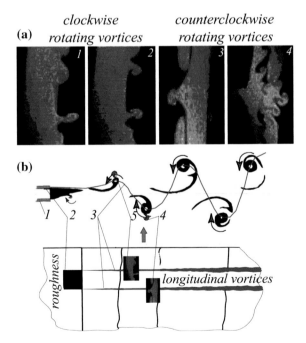

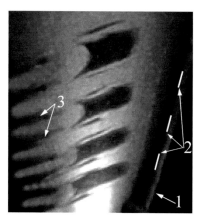

Fig. 6.12 Longitudinal structures of stationary disturbances generated by transversally spaced roughness elements: channel outlet (*1*), roughness elements (*2*) and a pair of longitudinal perturbations behind one the elements (*3*)

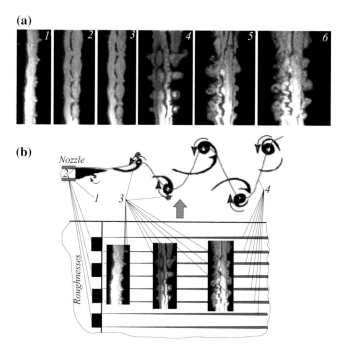

Fig. 6.13 Interaction of the forced longitudinal disturbances with the vortex street: perturbed vortices at increase of the streamwise distance (from *1* to *6*) (**a**); sketch of the vortex street with the exit of the extension channel (*1*), roughness elements (*2*), 3D bursts of the cylindrical vortices (*3*), and longitudinal structures (*4*) (**b**)

perturbations of both clockwise and counterclockwise rotating cylindrical vortices (images *1–4* in Fig. 6.11*a*). This is in contrast to the jet with a top-hat profile where the longitudinal structures excited by roughness of the nozzle surface affect just a thin shear layer.

Similarly, an array of roughness elements mounted in the extension channel (e.g., four of them as in Fig. 6.12) results in multiple longitudinal disturbances. Initially they induce a transversally periodic flow perturbation at one side of the vortex street corresponding to the roughness position, see image *1* in Fig. 6.13. Further downstream the entire street of cylindrical vortices becomes disturbed by three-dimensional bursts testifying to the laminar flow breakdown, see images *2–6*.

6.4 Dynamics of the Turbulent Plane Jet

While thickening in the extension channel, the boundary layers perturbed by the sand paper strip in the basic nozzle converge to a turbulent mean velocity profile at the jet origin as is shown in Fig. 6.14. Note transverse spreading of the turbulent jet starting from the extension channel outlet which is different from the behavior of laminar jets (compare to Fig. 6.3). In Fig. 6.15 one can observe that in natural conditions (image *1*) the spreading angle makes about forty degrees being more than twice larger than that of the laminar plane jet.

Figure 6.15 also demonstrates the flow response to external acoustic forcing at one and the same sound pressure level and different frequencies. Characteristics of the induced sinusoidal oscillations of the jet clearly shown by images *2–5* are summarized in Table 6.2. Within experimental accuracy, they follow similar results on the laminar jet oscillations presented in Table 6.1. The only difference is that at

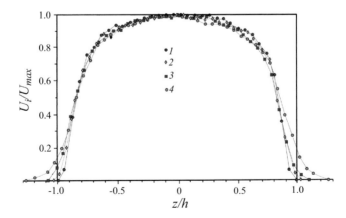

Fig. 6.14 Mean velocity profiles of the turbulent plane jet at $x = 2$ (*1*), 5 (*2*), 9 (*3*), and 15 mm (*4*) behind the extension channel outlet

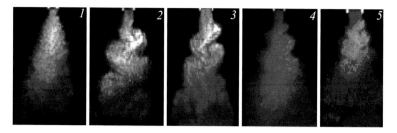

Fig. 6.15 Visualization of the turbulent plane jet in natural conditions (*1*) and under external acoustic forcing at frequencies f = 30 (*2*), 40 (*3*), 50 (*4*) and 60 Hz (*5*) (see Presentation Chap. 6: "Multimedia file No. 6.9") (http://extras.springer.com)

Table 6.2 Characteristics of the turbulent jet oscillations

Conditions	Frequency (f, Hz)	Wave length (λ, mm)	Propagation velocity ($c = f\lambda$, m/s)	Non-dimensional values	
				Sh = $2fh/U_0$	c/U_0
Natural, Fig. 6.14 (*1*)	–	–	–	–	–
Forced, Fig. 6.14 (*2–5*)	30	50	1.5	0.12	0.40
	40	40	1.6	0.16	0.43
	50	35	1.75	0.19	0.47
	60	29	1.7	0.23	0.46

flow visualization in natural conditions, the periodicity of turbulent jet is not observed. Most likely, this is due to high-amplitude background velocity disturbances overwhelming the coherent motion which is less pronounced in the absence of external forcing.

The above data testify to the sinusoidal oscillations of laminar and turbulent plane jets as a fairly universal phenomenon, at least, at their periodic excitation with appropriate frequencies. Large-scale anti-symmetric structure of flow perturbations in turbulent plane jets was also observed by other authors, e.g. Antonia et al. (1983).

6.5   Instabilities of Plane and Round Jets: A Comparison

Mean velocity profiles of the plane laminar and turbulent jets at the outlet of the extension channel are very similar to the velocity distributions in the axisymmetric flow, compare Figs. 6.16 and 4.9. However, dynamics of the jets originating in these configurations is quite different. Summarizing the experimental data on this point, discussed in the present and previous chapters, the following is to be emphasized.

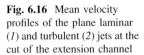

Fig. 6.16 Mean velocity
profiles of the plane laminar
(*1*) and turbulent (*2*) jets at the
cut of the extension channel

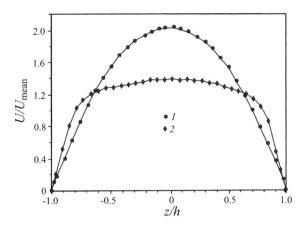

The laminar round jet with the initial parabolic mean velocity profile is distinguished by an extended laminar region without visible coherent disturbances. Otherwise, the laminar plane jet with similar velocity distribution is prone to anti-symmetric instability and experiences large-scale sinusoidal oscillations. If so, the influence of acoustic excitation upon the laminar part of the round jet is negligible while vortex structures of the plane jet can be effectively controlled by external periodic forcing. In the same way, the turbulent plane and round jets are different by sinusoidal perturbations of the first of them which are stimulated by harmonic excitation, in our case, by sound waves. The fact that transverse spreading of the turbulent plane jet is much faster than that of the round one seems relevant to such oscillatory behavior.

Concerning interactions of the longitudinal structures of perturbations with the coherent vortices, at first, we focus on the laminar round and plane jets with the top-hat mean velocity profile. In the round jet, the interaction results in three-dimensional bursts appearing as Λ- or Ω-like vortices spaced over the jet periphery and disturbing the entire circumference. In the plane configuration, the shear layers on both sides of the jet evolve independently of each other making it possible to modify them separately through excitation of controlled longitudinal perturbations. In the case of a parabolic mean velocity profile, the effect of longitudinal disturbances on the flow pattern in the laminar region of the round jet is probably small, if any. By contrast, laminar flow breakdown and jet mixing with surrounding air are obviously enhanced at the interaction of the longitudinal structures with the cylindrical vortices of the plane jet.

Key points

- Oscillatory behavior of the laminar plane jet with a parabolic mean velocity profile and that of the turbulent plane jet are very similar with large scale anti-symmetric oscillations dominating the flow pattern.

- Both laminar and turbulent plane jets are receptive to external acoustic excitation so that the characteristics of their oscillations can be modified at variation of the forcing frequency.
- Longitudinal structures of velocity perturbations generated by surface roughness at one side of the nozzle modulate the adjacent shear layer of the plane jet with a top-hat mean velocity profile and the entire flow pattern of the plane parabolic jet.

References

Antonia RA, Browne LWB, Rajagopalan S, Chambers AJ (1983) On the organized motion of a turbulent plane jet. J Fluid Mech 134:49–66

Boiko AV, Grek GR, Dovgal AV, Kozlov VV (2002) The origin of turbulence in near-wall flows. Springer, Berlin, pp 1–267

Boiko AV, Chun HH, Litvinenko MV, Kozlov VV, Cherednichenko EE, Lee I (2005) Longitudinal structures in plane jets. Dokl Phys 50(7):377–379

Boiko AV, Dovgal AV, Grek GR, Kozlov VV (2012) Physics of transitional shear flows. Springer, Dordrecht, pp 1–298

Hussain AKMF, Thompson CA (1980) Controlled symmetric perturbation of the plane jet: an experimental study in the initial region. J Fluid Mech 100:97–431

Kozlov VV, Grek GR, Löfdahl L, Chernorai VG, Litvinenko MV (2002) Role of localized streamwise structures in the process of transition to turbulence in boundary layers and jets (review). J Appl Mech Tech Phys 43(2):224–236

Sato H (1960) The stability and transition of a two-dimensional jet. J Fluid Mech 7(1):53–80

Schlichting H (1934) Laminare Kanaleinlaufstromung. ZAMM 14:368–373

Yu MH, Monkewitz PA (1990) The effect of nonuniform density on the absolute instability of two-dimensional inertial jets and wakes. Phys Fluids A 2:1175–1198

Yu MH, Monkewitz PA (1993) Oscillations in the near field of a heated two-dimensional jet. J Fluid Mech 255:323–347

Chapter 7
Round Jets in a Cross Shear Flow

In this chapter, results of experimental and numerical studies on characteristics of development of a round jet with a parabolic mean velocity profile at the nozzle exit in cross flow are presented. Basic distinctions in characteristics of development of a round jet with a top-hat and a parabolic mean velocity profiles at the nozzle exit are shown. It is revealed, that instability of a round jet with a parabolic mean velocity profile at the nozzle exit results in its deformation as tangential bursts of gas from the jet periphery in ambient space by means of a cross stream, folding them in a pair of counter rotating stationary vortices and hence reducing the jet core size. It is found, that the most unstable modes with high frequencies represent wave packets on a pair of counter rotating stationary vortices and instability modes at low frequencies develop in a jet wake closer to a wall. Growth of penetration of the jet in shear cross flow and a reduction in the near-field entrainment of cross flow fluid by a parabolic jet are found. It is shown, that the jet/cross flow interfaces of the parabolic jet might have undergone a "stretching and thinning" process caused by the cross flow. For more illustrations on this topic see also Chap. 7: "Multimedia files Nos. 7.1–7.5" (http://extras.springer.com).

7.1 Experimental Setup and Measurement Technique

The scheme and photo of the setup allowing us to change the initial conditions at the nozzle exit are shown in Figs. 7.1 and 7.2. A basis of the setup is a classical short nozzle (Vitoshinsky nozzle *1*) with a honeycomb (*2*) and a set of grids (*3*) in the settling chamber (*4*). Pipes (*5*) of different length could be joined to the classical nozzle exit. The internal diameter of the pipes was equal to inner diameter of the classical nozzle ($d = 20$ mm). The present configuration allowed us to change initial conditions of the jet at the nozzle exit, gradually forming a parabolic mean velocity

© The Author(s) 2016
V.V. Kozlov et al., *Visualization of Conventional and Combusting
Subsonic Jet Instabilities*, SpringerBriefs in Applied Sciences and Technology,
DOI 10.1007/978-3-319-26958-0_7

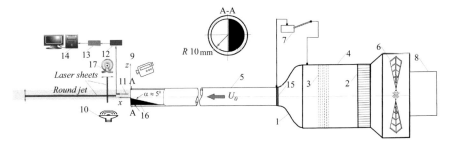

Fig. 7.1 Scheme of the round jet setup and measuring system: Vitoshinsky nozzle (*1*), honeycomb (*2*), grids (*3*), settling chamber (*4*), extension pipe of 430, 870, 1300, or 4000-mm length (*5*), fan (*6*), liquid micro-manometer (*7*), smoke generator (*8*), video camera (*9*), dynamic loudspeaker (*10*), hot-wire probe (*11*), hot-wire anemometer (*12*), analog-to-digital converter (*13*), computer (*14*), flow turbulizer (*15*), overlay (*16*), fan creating a weak cross flow (*17*)

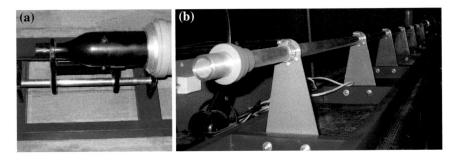

Fig. 7.2 Experimental setup with short (**a**) and long (**b**) nozzles

profile at the jet origin. The flow in the channel (*5*) was created using a fan (*6*), flow velocity at the nozzle exit making $U_0 = 5$ m/s ($Re = U_0 \times d/v = 6667$) was measured by a liquid micro-manometer (*7*).

Flow visualization and hot-wire measurements in the jet cross sections at different distances from the nozzle exit were performed. The visualization was carried out using a smoke generator of industrial production. The jet was seeded with smoke and illuminated by a narrow laser sheet, and jet evolution was fixed by a digital video camera both in a general view and in longitudinal and cross sections. The given technique was described in detail in the work by Litvinenko et al. (2004) where it was used to study a classical laminar jet development and its turbulization mechanism. In the mentioned work, as well as in the present case, influence of acoustic waves generated by a dynamic loudspeaker on the jet structure was investigated. It is necessary to note that the laser flashes were synchronized with the generator of pulses for the loudspeaker operation so that the jet evolution in a stroboscope mode could be visualized.

Hot-wire measurements in the jet were carried out using a DISA constant temperature anemometer. Both mean (*U*) and fluctuation (*u′*) streamwise velocity components were measured at different distances from the nozzle exit. The hot wire made of gilded tungsten was 1 mm in length and 5 μm in diameter. The anemometer probe (*11*) was calibrated in the free stream at an overheat ratio of 1.8 against a modified King's law: $U = k_1 (E^2 - E_0^2)^{1/n} + k_2 (E - E_0)^{1/2}$, where E and E_0—output voltages of the anemometer at flow velocity switched on and off, respectively, k_1, k_2 and n are constants. The factor n is usually close to 0.5, the second constant k_2 takes into account free convection at the wall at small flow velocities. The maximum error at the probe calibration did not exceed 1 % of U_0. The hot-wire signal was digitized by an analog-to-digital converter (*13*) with further data acquisition and processing in a personal computer (*14*). The jet was traversed by the probe with a step of 0.5 mm in three cross sections at $x = 2$, 10 and 20 mm from the nozzle exit. Downstream evolution of both a laminar and a turbulent jet was investigated. The turbulent jet was generated using a turbulizer (*15*) representing a sand paper which was 5 mm in width pasted on the inner surface of the short classical nozzle. The laminar and turbulent jets were examined at one and the same flow velocity $U_0 = 5$ m/s.

In the following graphs, the abscissa r is normalized as r/R where R is the radius of the nozzle exit. The ordinate is reduced by the mean value of the jet velocity U_{mean} and the velocity at the channel axis U_0 so that rms amplitudes of velocity pulsations u' are given as percentage of U_0.

7.2 Results of Measurements and Visualization

7.2.1 Round Jet with a Classical Top-Hat Mean Velocity Profile at the Nozzle Exit

A facility with a short classical nozzle designed to generate a round jet with a top-hat mean velocity profile at the nozzle exit and visualization data obtained in longitudinal and cross sections are shown in Fig. 7.3. It is possible to observe the presence of Kelvin–Helmholtz ring vortices and a result of their interaction with streaky structures generated by roughness elements pasted on the inner surface of the nozzle. Disturbances interaction results in azimuthal vortices appearing as Λ- or Ω-like structures. Exactly the dynamics of these structures and their evolution result in intensification of the jet mixing process with ambient air and, finally, in jet turbulization.

Mean (U) and fluctuation (u') velocity profiles measured at different distances from the nozzle exit are shown in Fig. 7.4. It is seen, that the mean velocity profile across the classical jet represents a very thin shear layer at the jet's periphery and an extended region without velocity gradient in the jet core. Such velocity distribution is referred to as a "top-hat" profile.

Apparently, the mean velocity distribution, distinguished by a strong gradient in the narrow region surrounding the jet core, brings about Kelvin–Helmholtz ring

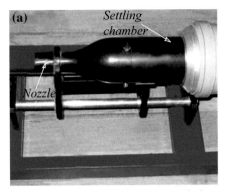

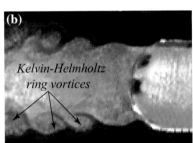

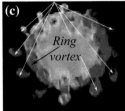

Fig. 7.3 Experimental setup with a short classical nozzle for generation of a round jet with a top-hat mean velocity profile at the nozzle exit (**a**), smoke visualization patterns of the round jet in the longitudinal (**b**) and cross (**c**) sections

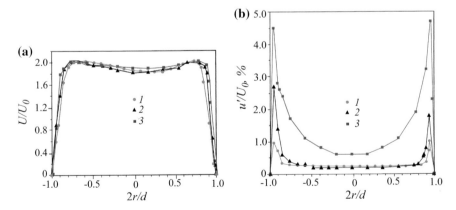

Fig. 7.4 Top-hat mean (**a**) and fluctuation (**b**) velocity profiles across the round jet at different distances from the nozzle exit (*1, 2,* and *3* correspond to $x = 5, 10,$ and 20 mm), $U_0 = 5$ m/s ($Re \approx U_0 \times d/v \approx 6700$)

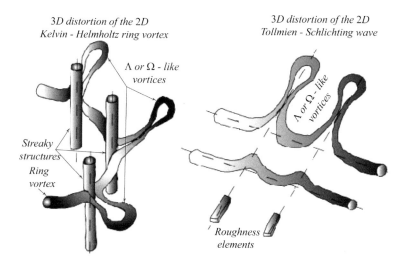

Fig. 7.5 Scenarios of interaction between streaky structures and ring vortex (*left*) and three-dimensional distortion of the Tollmien–Schlichting wave caused by roughness elements (*right*)

vortices generation. In this case, the maximum level of velocity perturbations observed in the region of the jet shear layer makes about 1 % of U_0 near the nozzle and grows up to 4.5 % of U_0 downstream. In the region of a jet core, the level of velocity disturbances is much lower when changing from 0.4 % of U_0 near the nozzle up to 1 % of U_0 downstream.

A scenario of the generation of azimuthal vortex structures such as Λ- or Ω-like vortices, through three-dimensional distortion of the ring vortices at the jet inhomogeneities induced by streaky structures (Fig. 7.5, *left*), has been offered (Litvinenko et al. 2004). This process looks similar to the classical scenario of three-dimensional distortion of a two-dimensional Tollmien–Schlichting wave at the nonlinear stage of boundary-layer transition or its deformation by roughness elements (Fig. 7.5, *right*). As a whole, the experimental data on the round jet with a top-hat mean velocity profile at the nozzle exit have shown the following:

- Streaky structures can be generated directly at the nozzle exit both under natural conditions and in conditions of their generation by roughness elements.
- The mechanism of interaction between the ring vortices and the streaky structures is nonlinear resembling the classical scenario of three-dimensional distortion of a two-dimensional Tollmien–Schlichting wave at the nonlinear stage of transition to turbulence or its modulation by roughness elements.
- Interaction of the disturbances results in occurrence of azimuthal vortex structures such as Λ- or Ω- like vortices stimulating mixing with ambient air and turbulization of the jet.

7.2.2 Round Jet with a Parabolic Mean Velocity Profile at the Nozzle Exit

Experimental facility with the channel length increased up to 4000 mm to model a round jet with a parabolic mean velocity profile at the nozzle exit (Fig. 7.6, *top*) and visualization of the jet in longitudinal and cross sections (Fig. 7.6, *bottom*) are presented in Fig. 7.6. An extended region of the laminar flow (about 200 mm— length) without pronounced ring vortices is observed in Fig. 7.6, *bottom*. Further downstream, transition to turbulence is found with regular bursts from the jet periphery appearing as helical structures (Fig. 7.6, *bottom*, sections *a–f*).

Experimental data on instability of the laminar round jet with a parabolic mean velocity profile are considered below. Note that in the laminar-flow region, the jet does not respond to external acoustic forcing, that is, vortex structures are not generated irrespective of frequency and intensity of the excitation. (On the contrary, in the round jet with a top-hat mean velocity profile subjected to acoustic oscillations, the streaky structures and their interaction with the ring vortices are stimulated.) The acoustic effects on the jet turbulization and mixing process are observed further downstream. In particular, the ring vortices are modified by the excitation depending on its frequency and amplitude (Kozlov et al. 2008).

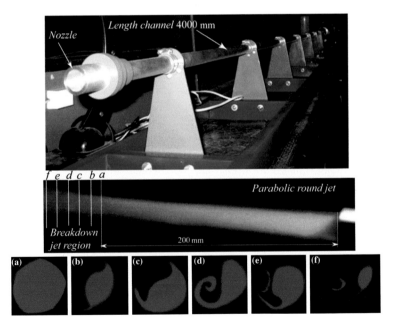

Fig. 7.6 Round-jet facility with the channel extended up to 4000 mm for generation of the parabolic mean velocity profile at the nozzle exit (*top*) and flow visualization in longitudinal (*bottom*) and cross (**a–f**) sections

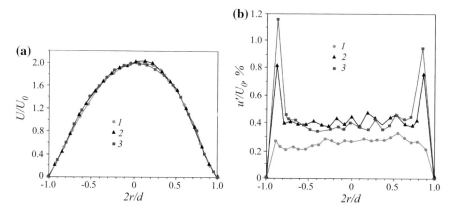

Fig. 7.7 Mean (**a**) and fluctuation (**b**) velocity profiles across the parabolic round jet at different distances from the nozzle exit (*1, 2,* and *3* correspond to $x = 2$, 10, and 20 mm), $U_0 = 5$ m/s ($Re \approx U_0 \times d/v \approx 6700$)

The mean (U) and fluctuation (u') velocity profiles across the laminar round jet at different distances from the nozzle exit are shown in Fig. 7.7. We see that the mean velocity profile at the nozzle exit has a parabolic shape representing, contrary to the round jet with a top-hat mean velocity profile at the nozzle exit, a continuous shear of flow velocity.

A maximum level of velocity pulsations is observed in the region of the jet shear layer and makes about 0.25 % of U_0 near the nozzle and grows up to 1.2 % of U_0 downstream. In the region of the jet core, the amplitude of velocity perturbations is much lower being approximately 0.4 % of U_0. As a whole, in comparison with the previous case, the level of velocity disturbances is obviously decreased in the region of jet periphery (from 4.5 % of U_0 up to 1.2 % of U_0 at $x = 20$ mm).

Thus, the basic difference between round jets with a top-hat and a parabolic mean velocity profile at the nozzle exit is that the latter is distinguished by an extended laminar-flow region without pronounced Kelvin–Helmholtz ring vortices. Also, this section of the jet does not respond to acoustic forcing with generation of vortex structures which could be detected.

7.2.3 Modeling Instability of a Round Jet with Parabolic Mean Velocity Profile to a Weak Cross Flow

Flow patterns in the jet cross sections shown in Fig. 7.6 resemble the structures found in a round jet in the presence of external cross flow (Lim et al. 2001). As it was noted by Lim et al. (2001), even at the top-hat mean velocity profile at the nozzle exit, distortion of the ring vortices under the influence of cross flow had no relation to the occurrence of vortex structures which they observed. Thus one expects that the

Fig. 7.8 Distribution of the
mean velocity of the round jet
at the nozzle exit without the
overlay (*1*) and with it (2),
flow velocity at an jet axis
$U_0 = 4$ m/s ($Re \approx U_0 \times d/$
$\nu \approx 5300$)

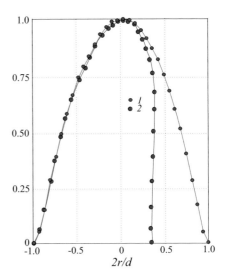

jet deformation and occurrence of large-scale vortex structures is connected to
instability of the jet to the cross flow resulting in distortion of a round jet with
initially parabolic mean velocity profile. For modeling of this phenomenon, an
experiment was carried out with a round jet deformed by an overlay (Fig. 7.1, *16*) to
reproduce the structures arising in natural conditions and subjected to a weak cross
flow generated by a fan (Fig. 7.1, *17*) (Grek et al. 2009). In this way, the mean
velocity profile at the nozzle exit was deformed close to the overlay, only (Fig. 7.8,
a curve 2). Mean velocity at the jet axis was $U_0 = 4$ m/s ($Re = U_0 \times d/\nu) \approx 5300$ and
the cross flow velocity was $U_\infty = 0.5$ m/s at their ratio $U_0/U_\infty = 8$; here *d* is the
nozzle diameter and *v* is air kinematic viscosity.

Typical visualization results demonstrating the instability of laminar round jet
with a parabolic mean velocity profile at the nozzle exit to a weak cross flow are
shown in Fig. 7.9. We see that the laminar jet is subjected to deformation as
"crescent" (Fig. 7.9, *left*) due to ejection of air from the jet periphery in two
tangential jets having the tendency to fold into two counter rotating vortices
(Fig. 7.9, *center* and *right*).

Cross
flow

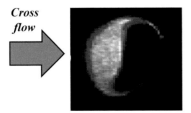 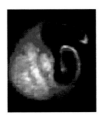 

Fig. 7.9 Visualization of the round jet subjected to a weak cross flow in cross sections at *x/d* = 2,
4, and 6 from the nozzle exit (*left-to-right*), see Presentation Chap. 7: "Multimedia file No. 7.1
(http://extras.springer.com)

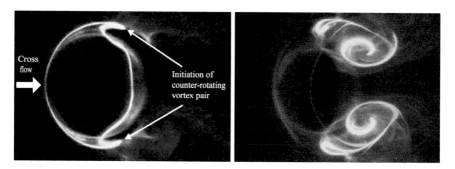

Fig. 7.10 Visualization of the jet with the laser plane perpendicular to the jet axis demonstrating the initiation of counter rotating vortex pair under the effect of cross flow (Lim et al. 2001)

The above flow patterns are qualitatively similar to those of a round jet with a top-hat mean velocity profile at the nozzle exit also affected by cross flow. [Fig. 7.10 is taken from the work by Lim et al. (2001)]. In this case, the jet-to-cross flow velocity ratio is $U_0/U_\infty = 4.6$.

It is distinctly visible in Fig. 7.10 that the cross flow results in tangential ejections from the jet in the region of its periphery and these ejections have characteristic arrow-like shape. They are symmetric to the cross flow velocity vector and have a tendency to fold into two symmetric counter rotating vortices. This process is not connected to the origination and evolution of Kelvin–Helmholtz ring vortices in the classical round jet. Apparently, the entire jet is involved in the deformation by the cross flow. Actually, one and the same process is observed in Figs. 7.9 and 7.10, except for certain asymmetry in shaping of two counter rotating vortices shown in Fig. 7.9. The latter is, probably, due to some non-parallelism of the cross flow velocity vector and the symmetry axis of the artificially deformed jet. A scheme of the folding of a round jet with a top-hat mean velocity profile at the nozzle exit into two counter rotating vortices under the effect of cross flow is given in Fig. 7.11.

Fig. 7.11 Folding of the round jet with a top-hat mean velocity profile at the nozzle exit into the counter-rotating vortex pair under the action of cross flow (Lim et al. 2001)

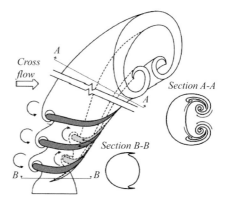

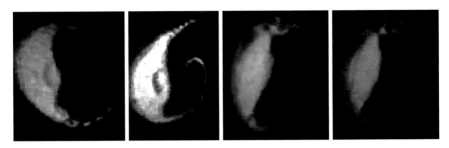

Fig. 7.12 Visualization of the round-jet cross sections with high-frequency secondary perturbations initiated by acoustics excitation (see Presentation Chap. 7: "Multimedia file No. 7.2") (http://extras.springer.com)

We see that the jet under influence of cross flow at the velocity ratio $U_0/U_\infty = 4.6$ is rotated opposite to the cross flow direction. In our case, the influence of a weak cross at $U_0/U_\infty = 8$ on the direction of the jet movement is practically not revealed. The scheme clearly displays the cross flow effect on the entire round jet. In section *A-A* two characteristic tangential ejections at the jet periphery are observed. Further downstream, in section *B-B*, the tangential ejections which are folded into the counter-rotating vortex pair proceed and the jet core becomes less pronounced. The same features are found in our experiment on the round jet breakdown under the effect of weak cross flow (Fig. 7.9).

When forcing the jet by external acoustic oscillations, secondary high-frequency breakdown of the jet in the regions of tangential ejections was observed. The flow patterns presented in Fig. 7.12 clearly show high-frequency disturbances in the regions of tangential ejections.

Thus, the present experimental studies on modeling of a laminar round jet with a parabolic mean velocity profile at the nozzle exit indicate that the instability of the jet to cross flow really exists; the cross flow results in deformation of the jet in the form of tangential ejections of air from the jet periphery in ambient fluid; due to the ejections, a pair of counter rotating vortices is generated and the jet core is diminished; the ejections are subjected to high-frequency secondary instability.

7.2.4 Round Jet in a Cross Shear Flow (in a Boundary Layer)

Consider the round jet evolution in a cross shear flow, that is, in a flat plate boundary layer. As it was already noticed in our Introduction, there is a great interest in studying jet flows from the viewpoint of their various technical applications. For example, a possibility of using jets as elements (actuators) of control systems in a boundary layer flow is considered. To solve such a problem, a deep understanding of the interaction between the jet and the cross shear flow is needed.

In the work by Selent (2008) the opportunity to control a turbulent boundary layer using the vortices generated by an actuator representing a jet blown into the near-wall flow is shown. In the work by Rist and Günes (2008) jet actuators for a shear flow control depending on frequency, blowing intensity and minor alterations of geometry and positions of the zones of optimum influence are investigated in detail.

Experimental data on instability of a laminar round jet with a parabolic mean velocity profile at the nozzle exit to a weak cross flow has shown that certain features of this instability can not be explained by well-known phenomena such as generation of Kelvin–Helmholtz vortices, excitation of varicose and sinusoidal modes of oscillations, etc. Generally, there are four main kinds of coherent structures typical of the jet in cross flow (see e.g. Fric and Roshko 1994; Kelso et al. 1996; Muppidi and Mahesh 2007). These are the counter rotating vortex pair, which originates in the near field of the jet and essentially follows the jet trajectory and dominates the flow field far downstream; shear-layer vortices which are located at the upstream side of the jet and take the form of ring-like or loop-like filaments; horseshoe vortices forming in the flat-plate boundary layer upstream of the jet exit and corresponding wall vortices downstream of the exit close to the wall; and "wake vortices/upright vortices" which are vertically oriented shedding vortices in the wake of the jet [Fig. 7.13 is taken from the work by Bagheri et al. (2009)].

A linear stability analysis shows that the jet in cross flow is characterized by self sustained global oscillations (Bagheri et al. 2009). A fully three dimensional unstable steady-state solution and its associated global eigenmodes are computed by direct numerical simulations. The steady flow, obtained by means of selective frequency damping, consists mainly of a (steady) counter rotating vortex pair in the far field and horseshoe-shaped vortices close to the wall. High-frequency unstable global eigenmodes are associated with shear-layer instabilities on the counter rotating vortex pair and low-frequency modes are associated with shedding vortices in the wake of the jet near the wall. In our case (see Sect. 7.2.3), probably, it is reasonable to speak only about instability of a jet in cross flow due to the high-frequency instability of a counter rotating vortex pair, because the cross flow is

Fig. 7.13 Direct numerical simulation of the round jet with parabolic mean velocity profile at the nozzle exit in cross flow [taken from the work by Bagheri et al. (2009)]

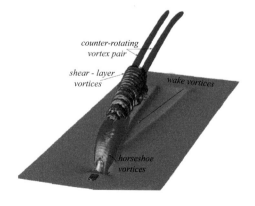

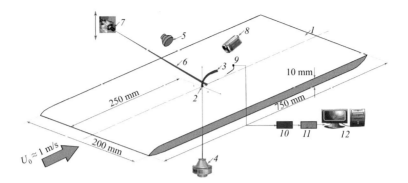

Fig. 7.14 Scheme of experiment: flat plate (*1*), orifice (*2*), round jet in a boundary layer (*3*), fan and smoke generator (*4*), dynamic loudspeaker (*5*), laser sheet (*6*), ruby laser (*7*), digital video camera (*8*), hot-wire probe (*9*), hot-wire anemometer (*10*), analog-to-digital converter (*11*), PC (*12*)

not a shear flow at the initial section of the jet as it has been shown by Bagheri et al. (2009) and New et al. (2006) where the jet was injected into a boundary layer.

Thus, one of the possible mechanisms of instability of a round laminar jet with a parabolic mean velocity profile at the nozzle exit in cross flow can be the mechanism of high-frequency instability of a counter rotating vortex pair (Bagheri et al. 2009). At least, secondary high-frequency instability of tangential ejections from a jet caused by cross flow is demonstrated in the present experiment.

Experimental studies were carried out on the development of a round jet with a parabolic mean velocity profile at the nozzle exit in a shear cross flow, that is, in a flat plate boundary layer. The objective was a continuation of studying of the jet structure and its downstream evolution in a cross flow, particularly in a shear flow, and also in comparison with the research data and results obtained by other authors (Bagheri et al. 2009; New et al. 2006).

The experimental scheme is shown in Fig. 7.14. A round jet (*3*) with a parabolic mean velocity profile at the nozzle exit was introduced in a flat plate boundary layer (*1*) through an orifice (*2*) of 5 mm diameter located at 250 mm from the flat plate leading edge. The parabolic mean velocity profile at the output of jet in the boundary layer was created at the exit of a long channel ($l/d = 200$, where l is the channel length and d is the intake orifice diameter). Flow visualization was performed using a smoke generator (*4*). The jet velocity at the orifice axis ($U_{jet} = 3$ m/s) reduced to the free stream velocity on flat plate ($U_0 \approx 1$ m/s) was $K = U_{jet}/U_0 \approx 3$ as in studies by Bagheri et al. (2009) and New et al. (2006).

The jet was forced by acoustic oscillations radiated by a dynamic loudspeaker (*5*) supplied with a sinusoidal signal varying its frequency and amplitude. Flow visualization was carried out by switching on a laser sheet (*6*) synchronized with the acoustic excitation. A general view of the jet and its cross sections were recorded by a digital video camera (*8*). Distributions of the mean (U) and fluctuation (u')

Fig. 7.15 Distribution of the mean (U_i) streamwise velocity component across the boundary layer at the point of jet input at $x = 250$ mm, $U_0 \approx 1$ m/s, $Re \approx U_0 \times x/ v \approx 16{,}700$ (y—actual distance, δ—boundary layer thickness)

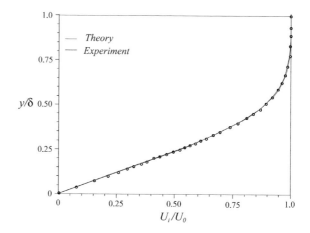

streamwise velocity components in the boundary layer and the jet were obtained through hot-wire measurements (*10*), digitizing of the data (*11*) and their further processing on PC (*12*).

Mean velocity profile $U_i/U_0 = f(y/\delta)$, measured at the point of jet injection in the boundary layer ($x = 250$ mm), is shown in Fig. 7.15. We see that the mean velocity profile $U_i/U_0 = f(y/\delta)$ corresponds to laminar flow in the boundary layer and is close to the Blasius distribution. The level of fluctuations intensity (u'/U_0) in a free stream is 0.17 % of U_0. Mean velocity profile ($U/U_{\max} = f(r/R)$) across the jet measured at the nozzle exit is shown in Fig. 7.16a. The latter distribution has a parabolic shape which is close to the mean velocity profile of a Poiseuille flow. The parabolic mean velocity profiles at the nozzle exit of a round jet injected into the boundary layer in the studies by Bagheri et al. (2009) and New et al. (2006) are shown here for comparison (see, Fig. 7.16b, c). A good qualitative agreement between the profile in Fig. 7.16a and the measurement data by New et al. (2006) for four values of $K = U_{\text{jet}}/U_0$ is found. Somewhat different is the distribution calculated by Bagheri et al. (2009) where, in the authors' words, "*The jet profile, mimicking the (laminar) parabolic velocity profile of pipe Poiseuille flow, is imposed as $v(r) = R(1 - r^2)$ exp $(-(r/0.7)^4)$ with r denoting the distance from the jet center, normalized by half the jet diameter D. Due to the super-Gaussian function, the profile has continuous derivatives for all r without a large modification of the parabolic shape near the jet centerline*".

Now we compare the flow patterns of round jets with parabolic mean velocity profiles at the nozzle exit introduced into the flat plate boundary layer in the present experiment and in the study by New et al. (2006). Qualitatively the visualization results are close to each other, see Fig. 7.17. The flow structure is observed in more detail in the right-hand image; however, the basic structure of the jet in cross flow is also seen in the left-hand one.

The initial part of the jet represents laminar flow without vortex structures that corresponds to the results on a laminar round jet with parabolic mean velocity

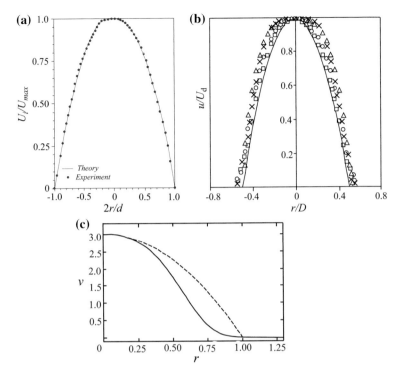

Fig. 7.16 Mean velocity profiles: present experiment (**a**); graph taken from New et al. (2006) (**b**); graph taken from Bagheri et al. (2009) (**c**); normal to the wall velocity component (v) is shown with *solid line* and standard Poiseuille parabolic mean velocity profile is shown with *dashed line*

profile at the nozzle exit (Kozlov et al. 2008). The jet core is gradually diminished by the increase in distance from the jet origin, while the jet-to-cross flow interface remains stable approximately up to its maximum bend. As it was noted by New et al. (2006), the parabolic jet shows strong stability to excitation of large-scale leading edge and lee-side vortices along shear layers in comparison with a similar jet with a top-hat mean velocity profile. In both cases of Fig. 7.17, large-scale vortex structures are observed further downstream. Thus, the basic core of an initially parabolic jet in cross flow is gradually rolled up into the counter rotating vortex pair. Instability of this vortex structure stimulates secondary high-frequency disturbances representing a ring or semicircle vortices (Bagheri et al. 2009).

Flow patterns in cross sections of the jet obtained through scanning it by a laser sheet parallel to the flat plate are presented in Fig. 7.18; in the top, visualization data by New et al. (2006) are also shown. The process of jet folding into the counter rotating vortex pair is demonstrated.

Close to the nozzle exit (Fig. 7.18, *1*) a body of the jet and a vortex sheet are observed. Further, a counter rotating vortex pair gradually forms (Fig. 7.18, *2–4*) and becomes distinctly seen in Image *5*.

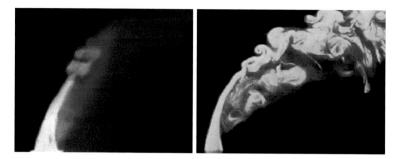

Fig. 7.17 Smoke visualization of the parabolic jet in a flat plate boundary layer at $K = U_{jet}/U_0 \approx 3$ in the present experiment (*left*) (see Presentation Chap. 7: "Multimedia file No. 7.3") (http://extras. springer.com) and flow pattern of the parabolic jet in a water channel boundary layer at $K = U_{jet}/U_0 \approx 3.5$ from the work by New et al. (2006) (*right*)

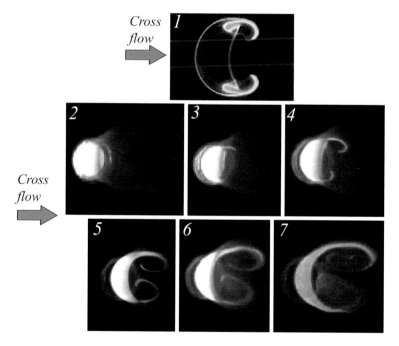

Fig. 7.18 Visualization of the jet cross sections: data by New et al. (2006) (*1*); results of the present experiment obtained at different distances from the nozzle exit in the wall-normal direction (*2–7*) (see Presentation Chap. 7: "Multimedia file No. 7.4") (http://extras.springer.com)

Visualization of the jet cross sections in the region where high-frequency perturbations evolve on the stationary counter rotating vortices is given in Fig. 7.19. Periodicity of the vortices due to acoustics effect at a frequency of 30 Hz is observed. As is found, Ω-like vortex structures develop separately on each of two

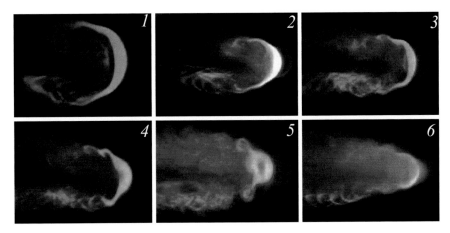

Fig. 7.19 Visualization of the jet cross sections at different distances from the nozzle exit in the wall-normal direction in the region of high-frequency secondary disturbances (*1–5*) and in the region of turbulent jet (*6*), acoustic waves are excited at a frequency of $f = 30$ Hz (see Presentation Chap. 7: "Multimedia file No. 7.5") (http://extras.springer.com)

Fig. 7.20 The most unstable
global modes with high
frequencies of the parabolic
round jet in a boundary layer
(Bagheri et al. 2009)

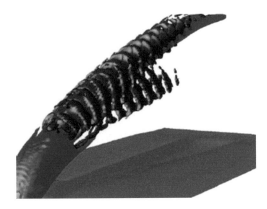

stationary counter rotating vortices. The latter observation is somewhat different from the numerical results by Bagheri et al. (2009) which testify to the high-frequency vortex structures surrounding both stationary vortices (Fig. 7.20).

The results of direct numerical simulation by Bagheri et al. (2009) on global instability of the jet in cross flow at $K = U_{jet}/U_0 = 3$ revealed the following features.

- The most unstable global modes with high frequencies are compact and represent localized wave packets on the counter rotating vortex pair. These modes are associated with the loop-shaped (Ω-like) vortex structures on the jet shear layer.
- The global modes with lower frequencies, on the other hand, also have a significant amplitude in the wake of the jet close to the wall and can be associated with less pronounced vortex structures arising downstream of the jet in the boundary layer.

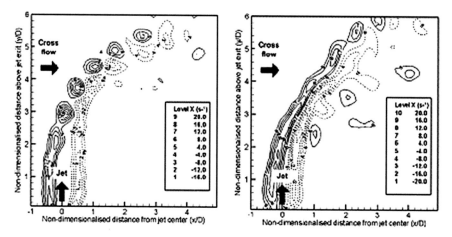

Fig. 7.21 Instantaneous vorticity fields along the symmetry plane for top-hat (*left*) and parabolic (*right*) jet in cross flow with $K \approx 3.5$ (New et al. 2006)

In the present experiment it was possible to visualize the formation of stationary counter rotating vortex pair with its instability to secondary high-frequency disturbances resulting in the wave packets appearing as Ω-like vortex structures. This observation is in agreement with the first conclusion from the study by Bagheri et al. (2009). However, we did not observe the low-frequency vortex structures in the jet wake close to the wall, likely because the structures in this region are about one order of magnitude smaller in amplitude than the structures on the counter rotating vortex pair as it is noted by Bagheri et al. (2009). In the present study only a vortex sheet was visible near the wall; probably, its propagation is connected to low-frequency vortex structures evolving in this flow region.

Detailed measurements of the instantaneous vorticity fields for the top-hat and parabolic round jets in cross flow were carried out in the work by New et al. (2006) with the data obtained at $K = U_{jet}/U_0 \approx 3.5$ shown in Fig. 7.21. Consistent with the flow visualization results, the top-hat round jet in cross flow produces distinct concentrated vortices in the vicinity of the jet exit along the jet/cross-flow interfaces (Fig. 7.21, *left*), whereas parabolic round jet in cross flow produces smaller-scaled vortices only after considerable distance downstream of the jet exit (Fig. 7.21, *right*).

The presence of undulations in the jet/cross-flow interfaces for parabolic round jet in cross flow indicates shear layer instabilities due to the mutual interaction between the jet and the cross-flow. The redistribution of shear layer vorticity is undoubtedly influenced by the parabolic velocity profile, with the thicker shear layer damping out possible perturbations and instabilities which would otherwise lead to the formation of large-scale vortices along the jet/cross flow interface regions. In contrast, the significantly thinner shear layer associated with the top-hat round jet in cross flow presents a much more favourable environment for large-scale vortices to form. Furthermore, the vorticity plots in Fig. 7.21 (*right*) indicate that the jet/cross flow interfaces of the parabolic round jet in cross flow might have undergone a "stretching and thinning" process caused by the cross flow, as can be inferred from the consistently higher vorticity levels at the interface compared to the top-hat round jet in cross flow (Fig. 7.21, *left*).

In the case of the top-hat jet in cross flow, vortex sheet stretching is limited by the earlier formation of the leading-edge and lee-side vortices (Fig. 7.21, *left*). On the other hand, the delay in the formation of the vortices in a parabolic jet in cross flow ensures that there is a greater streamwise distance for the stretching process to take place. From results of the vorticity measurements, New et al. (2006) draw a conclusion, that the near-field of the top-hat jet in cross flow has higher entrainment rate than the parabolic jet in cross flow due to the higher rate of formation of leading-edge and lee-side vortices. Also, their periodic formation ensures that the cross-flow fluid could be more readily entrained into the jet body, and this is consistent with the top-hat jet in cross flow's relatively lower penetration heights into the cross-flow. The mechanism of a liquid entrainment from a cross flow in a jet core becomes more clear, as presented in Fig. 7.22 that shows time-averaged streamlines along the symmetry plane for the top-hat and parabolic jet in cross flow. Despite the differences in the jet velocity profiles and the development of large-scale vortex structures, the overall streamline plots for both cases generally display similar topological structures with "unstable focus" (UF) features promi-nently at the lee- side regions. As for the "unstable focus", it has been previously observed by New et al. (2004) for an elliptic jet in cross flow, and also by Kelso et al. (1996), Hasselbrink and Mungal (2001) under significantly varied conditions, which they identified as an "unstable node". These critical points are linked to the manner in which the cross-flow recovers at the lee-side region of the jets.

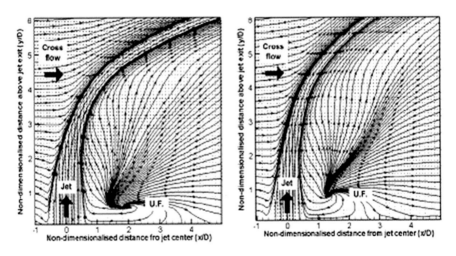

Fig. 7.22 Time-averaged streamlines along the symmetry plane for the top-hat (*left*) and parabolic (*right*) jet in cross flow with $K \approx 3.5$ (New et al. 2006). Unstable foci (*UF*) are observed at the lee-side of the jet both for the top-hat and parabolic jet in cross flow

We see from Fig. 7.22 that the contra flow is observed both for top-hat and parabolic jet in cross flow. There appears to be little variation in the location of the "unstable focus" (between $x/D = 1$ and 2) at the same K for either jet velocity profiles, although they always tend to form slightly further downstream (approximately 0.5 D) for a top-hat jet in cross flow. This suggests that the delayed formation of the leading-edge and lee-side vortices in the parabolic jet may have enabled the cross-flow to recover earlier or faster at the lee-side. Lastly, the vertical distances between the "unstable focus" and the tunnel floor were found to be between $y/D = 0.4$ and 1.0 for both cases. These results show that for a given K, the location of the "unstable focus" is not affected significantly by the jet velocity profile. However as a whole growth of jet penetration and a reduction in the near-field entrainment of cross-flow fluid by a parabolic jet in cross flow was observed.

Qualitative results of the present experiment do not provide answers to the processes happening in a near wall field of a parabolic round jet interacting with the boundary layer. Nevertheless, the importance of understanding the mechanism of interaction of a jet and a cross shear flow in this flow region is obvious (New et al. 2006) thus more quantitative measurements are being planned. Preliminary calculations of the process of interaction of a parabolic round jet with a cross shear flow have been carried out with the use of the "Fluent-program". At the first stage, as initial conditions for calculations, only mean velocity profiles of a jet, a cross flow, and a velocity ratio $K = U_{jet}/U_0 = 3$ are used. Naturally, it insufficient for detection of all features of flow interaction, nevertheless, it was possible to observe separate structural flow characteristics (Fig. 7.23). Calculation results well demonstrate the initial region of the jet where it is possible to observe the core thinning, vortex

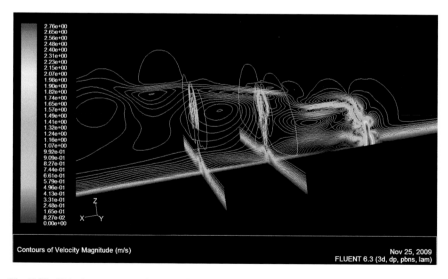

Fig. 7.23 Velocity contours of a round jet with parabolic mean velocity profile at the nozzle exit interacting with a flat plate boundary layer at $K = U_{jet}/U_0 = 3$ calculated using a Fluent software package

shedding in the near wall region at the lee-side and also existence of two counter rotating vortices in y-z planes. The isolines showing presence of large-scale vortices further downstream are, most likely, susceptible to initial and boundary conditions (acoustic environment, characteristics of the initial flow perturbations, etc.) which are not taken into account in these calculations. Nevertheless, the first attempt to use the Fluent programs for calculations of such flows seems useful.

As a whole, pilot studies on the interaction of a round jet with a parabolic mean velocity profile at the nozzle exit with a boundary layer indicate that more qualitative and quantitative data on the processes are required. For deep understanding of the interaction of a jet with cross flow it is necessary to examine round jets with top-hat and parabolic mean velocity profiles at the nozzle exit in more detail.

Key points

- Evolution of round jets with a top hat and a parabolic mean velocity profile at the nozzle exit is essentially different.
- Instability of a round jet with a parabolic mean velocity profile leads to its deformation in the form of tangential ejections from the jet periphery by means of cross flow into ambient space, folding of ejections into the counter rotating vortex pair and to reduction of the jet core size.
- A round jet with a parabolic mean velocity profile in a cross shear flow is subjected to folding into the counter rotating stationary vortex pair.
- The most unstable global modes with high frequencies represent wave packets on the counter rotating stationary vortex pair. These modes are connected with Ω-like vortex structures in the shear layer.
- Global modes at low frequencies also have considerable amplitude in the jet wake closer to the wall.
- Growth in jet penetration and reduction in near-field entrainment of cross-flow fluid by a parabolic jet in cross flow is found.
- The jet/crossflow interfaces of the parabolic jet in cross flow might have undergone a "stretching and thinning" process caused by the cross-flow.

References

Bagheri S, Schlatter Ph, Schmid PJ, Henningson DS (2009) Global stability of jet in crossflow. J Fluid Mech 624:33–44

Fric TF, Roshko A (1994) Vortical structure in the wake of a transverse jet. J Fluid Mech 279:1–47

Grek GR, Kozlov VV, Kozlov GV, Litvinenko YA (2009) Instability modeling of a laminar round jet with parabolic mean velocity profile at the nozzle exit. Byull. Novosibirsk State Univ., Ser. Fiz. 4(1):14–324, (in Russian)

Hasselbrink EF, Mungal MG (2001) Transverse jets and jet flames. Part 1. Scaling laws for strong transverse jets. J Fluid Mech 443:1–25

Kelso R, Lim T, Perry A (1996) An experimental study of round jets in cross-flow. J Fluid Mech 306:111–144

Kozlov GV, Grek GR, Sorokin AM, Litvinenko YA (2008) Influence of the initial conditions at the nozzle exit on a flow structure and stability of the plane jet. Byull. Novosibirsk State Univ. Ser. Fiz. 3(3):25–37 (in Russian)

Lim TT, New TH, Luo SC (2001) On the development of large scale structures of a jet normal to a cross-flow. Phys Fluids 3:770–775

Litvinenko MV, Kozlov VV, Kozlov GV, Grek GR (2004) Effect of streamwise streaky structures on turbulization of a circular jet. J Appl Mech Tech Phys 45(3):349–357

Muppidi S, Mahesh K (2007) Direct numerical simulation of round turbulent jets in crossflow. J Fluid Mech 574:59–84

New TH, Lim TT, Luo SC (2004) A flow field study of an elliptic jet in cross-flow using DPIV technique. Exp Fluids 36: 604–618

New TH, Lim TT, Luo SC (2006) Effects of jet velocity profiles on a round jet in cross-flow. Exp Fluids 40(3):859–875

Rist U, Günes H (2008) Qualitative and quantitative characterization of a jet and vortex actuator. Abstracts of 7th ERCOFTAC SIG33 Workshop, pp 35–35, Genova, Italy, 16–18 Oct 2008

Selent B (2008) DNS of jet in crossflow on a flat plate boundary layer. Abstracts of 7th ERCOFTAC SIG33 Workshop pp 31–31, Genova, Italy, 16–18 Oct 2008

Chapter 8
Subsonic Round and Plane
Macro- and Microjet in a Transverse
Acoustic Field

In this chapter we discuss results of experimental studies on evolution of round and plane macro- and microjets at small Reynolds numbers in a transverse acoustic field. Hot-wire measurements and smoke visualization of the jets using stroboscopic laser illumination synchronized with acoustic forcing made it possible to obtain new data on their development. Additional material to this chapter is given in the multimedia files, which supplement the book, titled Chap. 8: "Multimedia files Nos. 8.1−8.22" (http://extras.springer.com).

8.1 Round Macro- and Microjets

Round macrojets with a top-hat mean velocity profile at the nozzle exit (Fig. 8.1) are subjected to Kelvin–Helmholtz instability and origination of ring vortices resulting in flow turbulization (Kozlov et al. 2008a, b, 2009). Flow visualization shows that Kelvin–Helmholtz ring vortices interact with streaky structures which can be generated at the nozzle exit, thus producing three-dimensional azimuthal perturbations such as Λ- or Ω-like vortices. Downstream evolution of these disturbances results in intensification of the mixing process and, finally, to turbulization of the jet (Litvinenko et al. 2004).

Variation of initial conditions at the nozzle exit, i.e. distributions of mean and fluctuation velocity components in the jet cross section, has a pronounced effect on the nozzle's exit (Fig. 8.2) results in a laminar flow over a large distance up to $l/d = 10$, where l is the length of laminar region and d is the nozzle exit diameter, without Kelvin–Helmholtz ring vortices (Kozlov et al. 2008a, b). Acoustic effect on the laminar region of a parabolic round jet was not found at variations of the forcing frequency and amplitude in wide ranges. Further downstream, the jet becomes turbulent through amplification of vortex structures observed by other authors studying a round jet in transverse flow (Lim et al. 2001).

© The Author(s) 2016
V.V. Kozlov et al., *Visualization of Conventional and Combusting
Subsonic Jet Instabilities*, SpringerBriefs in Applied Sciences and Technology,
DOI 10.1007/978-3-319-26958-0_8

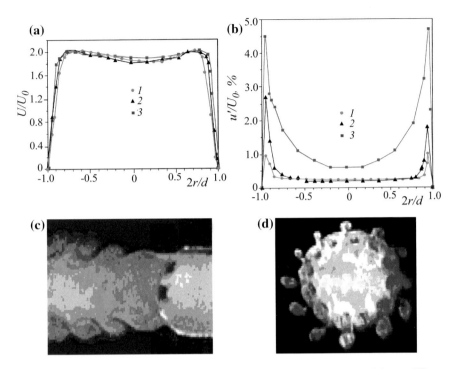

Fig. 8.1 Mean (**a**) and fluctuation (**b**) velocity profiles of the classical round jet at different distances from the nozzle exit (*1, 2, 3*—at x = 2, 10, 20 mm, respectively) and flow patterns of the jet in its streamwise (**c**) and cross (**d**) sections, U_0 = 5 m/s (Re = $U_0 \times d/v \approx 6700$)

Instability of a laminar jet with a parabolic mean velocity profile at the nozzle exit to weak cross flow was modeled in experiments by (Grek et al. 2009). Visualization results shown in Fig. 8.3 practically coincide with the research data by Lim et al. (2001). Tangential ejections of air from the jet periphery in ambient fluid, their folding into the counter rotating vortex pair and reduction of the jet core are clearly seen.

Also, experiments have shown that the Kelvin–Helmholtz instability of round jets with top-hat mean that the velocity profile at the nozzle exit is still found to be a reduction of the jet diameter to rather small values (Fig. 8.4). However, for a jet diameter less than about 1 mm, the jet core degenerates and the top-hat mean velocity profile turns into a parabolic one.

As a result of these previous studies, details on subsonic round macrojet developments have been revealed at variation of initial conditions at the nozzle exit including controlled forcing of the jet by acoustic oscillations. On this basis, round microjets with a parabolic and top-hat mean velocity profiles in the presence of transverse acoustic waves were investigated by the present authors. During these

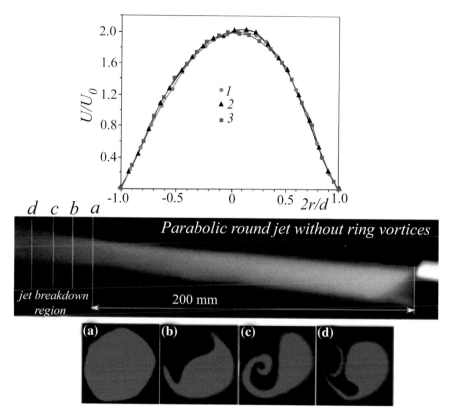

Fig. 8.2 Mean velocity profile (*top*) of the parabolic round jet at different distances from the nozzle exit and visualization of the jet in streamwise and cross sections (**a–d**), $U_0 = 5$ m/s (Re $= U_0 \times d/\nu \approx 6700$)

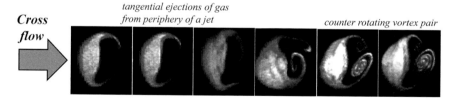

Fig. 8.3 Smoke visualization of the parabolic round jet cross sections subjected to weak cross flow

experiments it was found that, in the absence of acoustic excitation, the round microjet has an extended region of laminar flow irrespective of a particular mean velocity profile at the nozzle exit (Fig. 8.5).

It is necessary to note that, contrary to a round macrojet with a top-hat mean velocity profile at the nozzle exit, the Kelvin–Helmholtz instability is not revealed

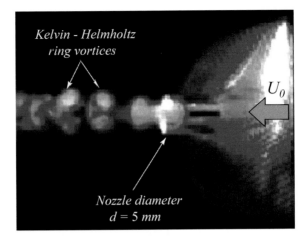

Fig. 8.4 Visualization of the laminar round jet with a top-hat mean velocity profile at the nozzle exit (U_0 = 1.5 m/s, d = 5 mm, Re_d =.500) (see Presentation Chap. 8: "Multimedia file No. 8.1") (http://extras.springer.com)

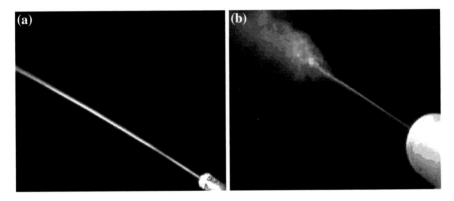

Fig. 8.5 Visualization of the round microjet without controlled acoustic forcing, parabolic (**a**) and top-hat (**b**) mean velocity profiles at the nozzle exit of d = 500 µm, Re_d = 50 (see Presentation Chap. 8: "Multimedia file No. 8.2") (http://extras.springer.com)

in this case. The velocity range of the round microjet in this experiment was from U_0 = 1.5 up to 10 m/s. Acoustic influence on the jet was created using a dynamic loudspeaker generating a sinusoidal instability mode. Frequency of the generated disturbances varied from 30 up to 1500 Hz at sound pressure level about 90 dB. As it is practically impossible to apply hot-wire technique for measurement of a mean velocity profile in a microjet, the top-hat and parabolic mean velocity profiles at the nozzle exit were distinguished according to the ratio l/d after the study by Kozlov et al. (2008a, b). On the other hand, there is an assumption that in a microjet the parabolic mean velocity profile at the micro-nozzle exit is always realized because

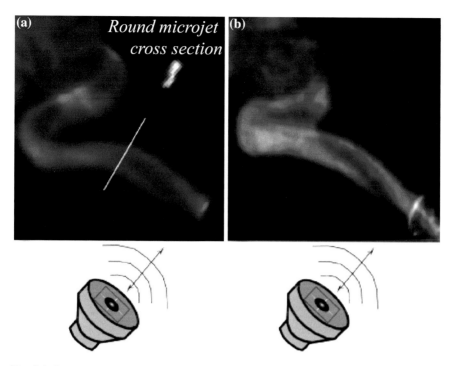

Fig. 8.6 Smoke visualization of the round microjet and its flattening in the transverse acoustic field ($f = 40$ Hz (**a**) and 100 Hz (**b**), sound pressure level is 90 dB) (see Presentation Chap. 8: "Multimedia files Nos. 8.2–8.4") (http://extras.springer.com)

the jet core practically disappears. Nevertheless, experiments with nozzles of 200, 400, 500, 600 and 1600-µm diameter have been carried out at a ratio $l/d \approx 50$ for realization of a parabolic mean velocity profile at the nozzle exit with gradual variation of the initial conditions. One expects that the flow emanating from the 1600-µm nozzle can hardly be referred to as a microjet, however, our preliminary data showed that the jets generated by 200, 400, 600 and 1600-µm nozzles are much similar.

Visualization of the round microjet emanating from the nozzle of 1600-µm diameter with velocity of 1.5 m/s ($Re_d = U_0 \times d/v = 60$) is shown in Fig. 8.6. It is seen that, under the influence of transverse acoustic waves (sound pressure level up to 90 dB), the round jet is deformed and transformed into the plane microjet. In this case, we assume that the effect of acoustic excitation is similar to that of the cross flow. Flattening of a round microjet, i.e. its transformation into the plane microjet due to transverse acoustic oscillations was also marked in other studies (Carpenter et al. 2009; Abramov et al. 1987; Hoover et al. 1991; Heister et al. 1997). It appears that the influence of transverse acoustic excitation on microjets is much more pronounced than that on macrojets. Acoustic forcing of a macrojet modifies the periodicity of vortex formation stimulating transition to turbulence, only.

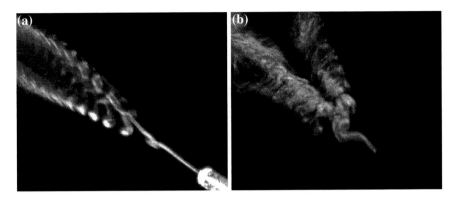

Fig. 8.7 Visualization of the round microjet with parabolic (**a**) and top-hat (**b**) mean velocity profiles at the nozzle exit; the nozzle diameter is 500 (**a**) and 1600 (**b**) μm, $Re_d = 50$ (**a**) and 160 (**b**) (see Presentation Chap. 8: "Multimedia files No. 8.3 and 8.4") (http://extras.springer.com)

Otherwise, the entire structure of a microjet becomes different with its bifurcation illustrated in Fig. 8.7. After the bifurcation, two jets evolve independently of each other with Ω-like vortex structures of secondary high-frequency perturbations. Repetition rate of the vortex structures depends on the frequency of acoustic forcing.

At the initial stage of breakdown, sinusoidal oscillations of the entire microjet are found. In both cases of parabolic and top-hat mean velocity profiles at the nozzle exit, the splitting of original jet under the influence of transverse acoustic waves is observed with an angle between secondary jets making about 20–25°. This effect takes place at low and high frequencies of acoustic excitation (ranging from 100 to 1500 Hz) and at different nozzle diameter (200, 400, 500, and 1600 μm), Fig. 8.8.

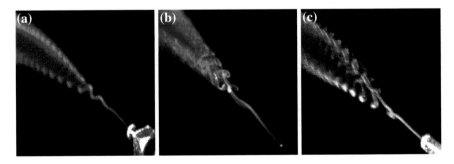

Fig. 8.8 Visualization of the round microjet evolution under the effect of transverse acoustic excitation ($f = 200$ Hz) at different nozzle diameter: 200 μm (**a**), 400 μm (**b**), and 500 μm (**c**); $Re_d = 20$ (**a**) 40 (**b**) and 50 (**c**) (see Presentation Chap. 8: "Multimedia files Nos. 8.5–8.7") (http://extras.springer.com)

The phenomenon considered here was not observed at a plane macrojet development (Kozlov et al. 2009), moreover, the mechanism of this process is not absolutely clear. An attempt to understand it and to suggest a hypothesis of its occurrence has been undertaken. A flow pattern of the round microjet transformed into the plane one, under the influence of transverse acoustic excitation, is represented in a large scale for clarity in Fig. 8.6. One can see that at the nozzle exit the round microjet is transformed into the plane microjet with sinusoidal oscillations in the direction of the acoustic velocity vector. Also, it is possible to observe a helical turning of the jet. Under the excitation, the sinusoidal oscillations rapidly grow downstream in the plane parallel to the direction of the acoustic velocity vector. This process results in splitting of the jet into two jets developing independently of each other. Both of them are prone to secondary high-frequency disturbances induced by acoustic forcing (see Figs. 8.7 and 8.8). Basically, the evolution of a microjet does not depend on the nozzle diameter (see Fig. 8.8) or the frequency of acoustic forcing (Fig. 8.9).

The above features of the round microjet are much different from those of round and plane macrojets, and were observed for the first time in the present studies. As a whole, the experimental data indicated that the Kelvin–Helmholtz instability of a

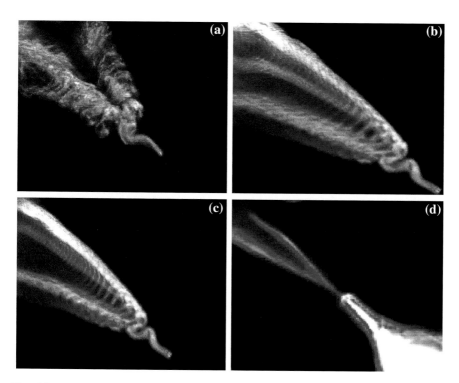

Fig. 8.9 Flow patterns of the round microjet forced by transverse acoustic oscillations at frequencies $f = 40$ (**a**), 150 (**b**), 300 (**c**) and 1500 (**d**) Hz; nozzle diameter, $d = 500$ μm (see Presentation Chap. 8: "Multimedia files Nos. 8.5–8.7") (http://extras.springer.com)

round jet with a top-hat mean velocity profile at the nozzle exit is still found when reducing the jet diameter to rather small values (see, Fig. 8.4). However at a jet diameter less than 1 mm, the top-hat mean velocity profile at the nozzle exit becomes almost parabolic due to disappearance of a non-gradient jet core and formation of a shear layer across the jet cross section. As it was already shown, the mechanism of the microjet evolution thus cardinally varies. The influence of transverse acoustic waves on a microjet appears considerably stronger than on the macrojet. The acoustic effects on a macrojet are reduced to the variation of periodicity of vortex formation and acceleration of its turbulization, while the microjet becomes completely different by transforming its round configuration into the plane one. Also at the splitting of a microjet, two secondary jets are generated with Ω-like vortex structures of high-frequency perturbations. The repetition rate of the Ω-like vortex structures depends on frequency of transverse acoustic excitation. Finally, we note that the details on the influence of transverse acoustic forcing on a microjet could be obtained with a nozzle diameter of 1.6 mm, assuming that the behavior of jets emanating from smaller nozzles was qualitatively the same.

8.2   Plane Macro- and Microjets

It is known that instability of a laminar plane jet is connected to occurrence and development of two instability modes: symmetric and asymmetric (Sato 1960) which are often associated with the concept of varicose and sinusoidal instabilities (Boiko et al. 2006). In the case of a top-hat mean velocity profile at the nozzle exit, these two modes compete with each other. At the initial stage of the jet evolution the symmetric instability dominates, then during gradual transformation of the top-hat mean velocity profile into the parabolic shape, the asymmetric mode becomes prevailing. In the experiments by Kozlov et al. (2008a, b, 2009) the parabolic mean velocity profile was generated immediately at the nozzle exit; in such case, only the asymmetric disturbances are expected. Visualization of a plane laminar macrojet with a parabolic mean velocity profile indicated that, in the absence of acoustic excitation, the whole jet was subjected to sinusoidal motion. For the first time, the existence of self-sustaining strong sinusoidal fluctuations of the plane macrojet with certain features of absolute instability was noted by Yu and Monkewitz (1993). The sinusoidal oscillations of a plane macrojet caused by its absolute instability in conditions of a parabolic mean velocity profile at the nozzle exit are shown in Fig. 8.10.

Some features of the jet evolution subjected to acoustic oscillations become clear at variation of the excitation frequency. At low-frequency forcing (30–70 Hz), the formation of asymmetric vortex structures is promoted, the angle of the jet spreading is increased from 18 to 30 degrees, and the transition to turbulence is accelerated. With growth of the excitation frequency from 90 to 150 Hz, the sinusoidal oscillations are suppressed and the jet splits into two jets. These observations support a conclusion by Sato (1960) that the plane jet with a parabolic

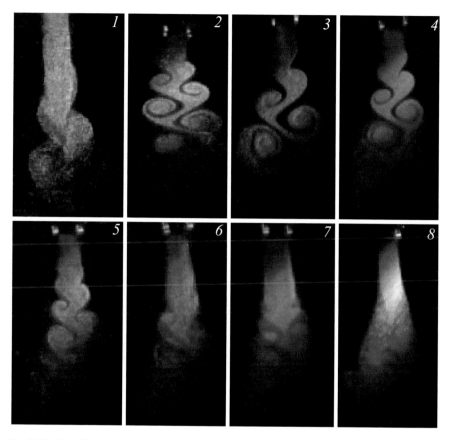

Fig. 8.10 Visualization of the laminar plane macrojet without acoustic effect (*1*) and at acoustic forcing (*2, 3, 4, 5, 6, 7,* and *8* correspond to f = 30, 40, 50, 60, 70, 90 and 150 Hz, respectively), U_0 = 3.7 m/s (Re$_h$ ≈ U_0 × h/v ≈ 3600)

mean velocity profile is subjected to instability of sinusoidal type and is more sensitive to low-frequency disturbances. When the nozzle width is reduced from 10 to 2.5 mm, the jet oscillations with formation of a sinusoidal vortex street are also observed (Fig. 8.11).

Results of the first qualitative experimental studies of the plane microjet (Kozlov et al. 2009) generated at the short classical nozzle exit of the width h = 700 μm showed that the jet oscillation with formation of a sinusoidal vortex street is actually the same as it was in the previous experiments (Fig. 8.12). In this case, the oscillation period is affected by acoustic perturbations, as well.

In the work by Gau et al. (2009) plane microjets (h = 50, 100, and 200 μm) were investigated at the length of nozzle slot l = 2000 μm in the absence of acoustic excitation. It was concluded that the microjet evolves without vortex formation and merging processes, thus being completely different from the macrojet flow. Interaction between the microjet and the ambient fluid has not been found before

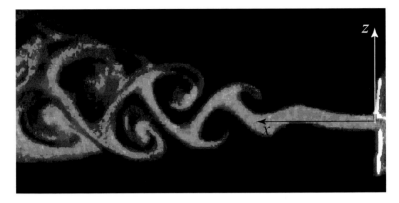

Fig. 8.11 Visualization of the laminar plane macrojet emanating from the nozzle of 2.5-mm width (jet is forced by transverse acoustic waves at 100-Hz frequency) (see Presentation Chap. 8: "Multimedia file No. 8.8") (http://extras.springer.com)

Fig. 8.12 Visualization of the laminar plane microjet generated at the nozzle of 700-μm width; the jet is forced by transverse acoustic waves at different frequencies: $f = 30$ (**a**), 80 (**b**) and 180 Hz (**c**)

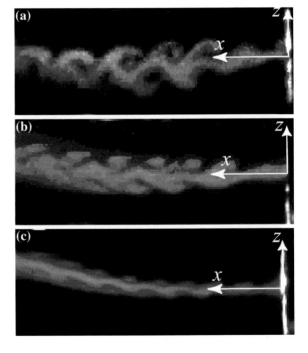

the jet breakdown that can occur at a higher Reynolds number at a location where rapid lateral expansion and mixing of the jet are observed.

Similar to the study by Gau et al. (2009), experiments have been performed with the plane microjet generated at the nozzle width $h = 200$ μm and the length of the nozzle slot $l = 2360$ μm ($l/h \approx 12$, Re = 100). A comparison of the visualization data

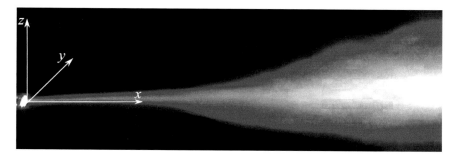

Fig. 8.13 Visualization of the laminar pseudo plane microjet without acoustic excitation: $h = 200$ μm, $Re_h = 100$

with the results of Gau et al. (2009) for the plane microjet emanating from the nozzle of $h = 200$ μm and $l = 2000$ μm ($l/h \approx 10$, $Re = 160$) was carried out. At small enough ratio $l/h \approx 10$, the jet can hardly be identified as a plane one. It seems better to considered it as an elliptic jet. In both cases, the pseudo plane microjet develops similarly in the absence of acoustic forcing, that is, one can observe laminar flow without sinusoidal oscillations with its turbulent breakdown further downstream (Fig. 8.13 and see Fig. 4 from the work by Gau et al. (2009)).

Significant variations of the pseudo plane microjet evolution can be observed at acoustic forcing. One can see in Fig. 8.14 that, irrespective of the acoustic velocity vector, the microjet splits into two jets developing independently of each other. This observation resembles the bifurcation of a round microjet under acoustic excitation that was discussed before. In certain conditions of acoustic frequency and jet velocity, it is possible to observe not only the jet bifurcation but also its trebling and even quadrupling (Fig. 8.15).

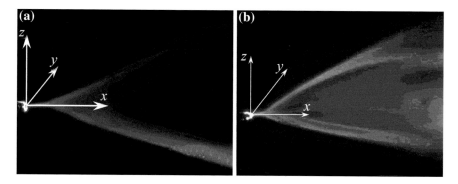

Fig. 8.14 Visualization of the laminar pseudo plane microjet at acoustic excitation ($f = 1700$ Hz), velocity vector of the acoustic oscillations is perpendicular (**a**) and parallel (**b**) to the jet plane; nozzle dimensions are $h = 200$ μm and $l = 2360$ μm (see Presentation Chap. 8: "Multimedia file No. 8.10") (http://extras.springer.com)

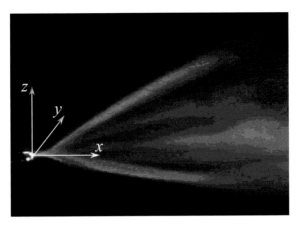

Fig. 8.15 Visualization of the laminar pseudo plane microjet at acoustic forcing ($f = 2500$ Hz); nozzle dimensions are $h = 200$ μm and $l = 2360$ μm

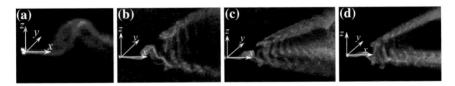

Fig. 8.16 Visualization of the laminar pseudo plane microjet at the acoustic effect (**a, b, c, d** stand for $f = 50$, 66, 80, 200 Hz, respectively)

In conditions of low-frequency acoustic excitation, besides the splitting of microjet, it is possible to observe several periods of its sinusoidal oscillations and vortex structures in the region of the secondary jets evolution (Fig. 8.16). Then, Fig. 8.17 shows that visualization data depend strongly on the jet section which is illuminated.

Thus, in experimental condition close to those of (Gau et al. 2009) it is found that acoustic excitation of a pseudo plane microjet results in its bifurcation and sinusoidal oscillation with the generation of vortex structures that was earlier observed at acoustic forcing of the plane macrojet (Kozlov et al. 2008a, b) and the round microjet (see Sect. 8.1).

Now we consider experimental results on a really plane microjet ($l/h = 70$, 180 at $l = 36{,}000$ μm and $h = 500$, 200 μm) with and without acoustic effects. In the absence of acoustic waves the plane microjet is subjected to sinusoidal oscillations similarly to the plane macrojet (Kozlov et al. 2008a, b, 2009). At reduction of the cross size of the nozzle slot from 500 to 200 μm, some features of the plane microjet were revealed.

In the absence of acoustic excitation the entire microjet is subjected to sinusoidal oscillations (Fig. 8.18). At acoustic influence upon the plane macrojet, its

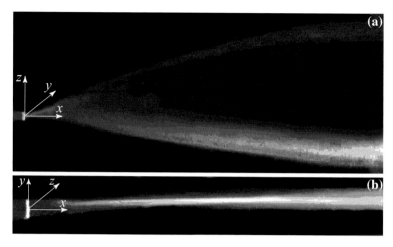

Fig. 8.17 Visualization of the laminar pseudo plane microjet at the acoustic forcing ($f = 1300$ Hz,): streamwise sections of the jet in a plane of the jet width h, (**a**) and in a plane of the jet length l, (**b**) (see Presentation Chap. 8: "Multimedia files No. 8.11 and 8.12") (http://extras.springer.com)

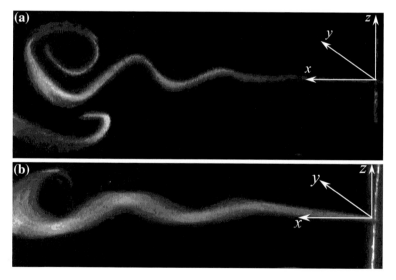

Fig. 8.18 Visualization of the laminar plane microjet without acoustic effect: nozzle slot $h = 500$ μm (**a**) and $h = 200$ μm (**b**) (see Presentation Chap. 8: "Multimedia file No. 8.13") (http://extras.springer.com)

oscillations generate a sinusoidal vortex street (Kozlov et al. 2008a, b, 2009). At low-frequency (30–150 Hz) transverse acoustic forcing of a microjet, alongside with occurrence of a sinusoidal vortex street it was possible to observe folding of

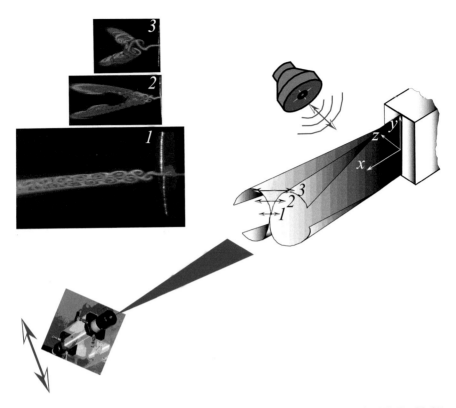

Fig. 8.19 Scheme of the plane microjet ($h = 200$ μm) under acoustic forcing ($f = 150$ Hz, 90 dB) and flow patterns in x-z planes at variation of the y coordinate (*1*, *2*, and *3* correspond to $y = 0$, 15, and 18 mm, respectively) (see Presentation Chap. 8: "Multimedia files No. 8.14 and 8.15") (http://extras.springer.com)

the jet at its edges in the direction of variable velocity vector of the flow created by acoustic waves. It is clearly seen in Fig. 8.19 that at sound pressure level of 90 dB the vortex sheet is rolled up in directions opposite to each other on each half-cycle of acoustic oscillations. The flow patterns of Fig. 8.19 demonstrate the sinusoidal vortex street in the center of the jet (Sect. 1, $y = 0$ mm). Also it is possible to observe a bifurcation of the plane jet (Sects. 2 and 3, $y = 15$, 18 mm). In more detail, the effect of transverse acoustic forcing on the initial portion of the plane microjet can be observed in Fig. 8.20.

Flow patterns of the plane microjet are shown in Fig. 8.21 where the sinusoidal vortex street is clearly seen in an x-z plane at $y = 0$, that is, in the jet centerplane. At nonzero y coordinates, the microjet splitting due to the oscillatory process caused by the transverse acoustic waves is visualized. A similar phenomenon was observed earlier when we studied the round microjet affected by transverse acoustic oscillations.

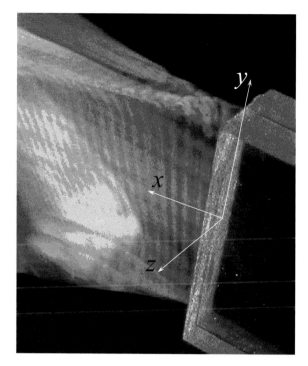

Fig. 8.20 Visualization of the plane microjet ($h = 200$ μm, $l = \pm18$ mm) forced by transverse acoustic oscillations at $f = 100$ Hz at sound pressure level of 90 dB

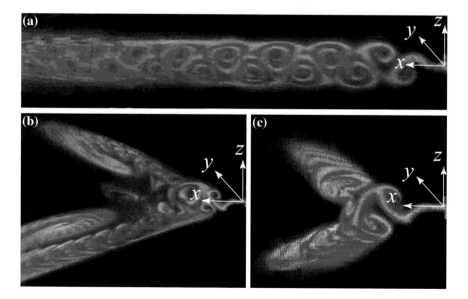

Fig. 8.21 Visualization of the plane microjet ($h = 200$ μm, $l = \pm18$ mm) under transverse acoustic excitation ($f = 100$ Hz, 90 dB) in x-z plane at different y—coordinates: $y = 0$ mm (**a**), ±15 mm (**b**) and ±18 mm (**c**)

The microjet bifurcation due to the periodic folding of its plane edges makes a new kind of plane microjet instability which is connected to comparability of microjet and acoustic energy. The same behavior was observed at the nozzle width $h = 500$ μm. The folding process is further illustrated in Fig. 8.22a. A sinusoidal vortex street extending far downstream is shown in Fig. 8.22b. Also, the folding process results in microjet extinction and its turbulization (Fig. 8.23).

Some results of particle image velocimetry of the plane microjet are presented in Presentation Chap. 8: "Multimedia files Nos. 8.18–8.20" (http://extras.springer.com). Effects of acoustic forcing and nozzle vibration on the plane microjet are compared in Presentation Chap. 8: "Multimedia files No. 8.21 and 8.22" (http://extras.springer. com).

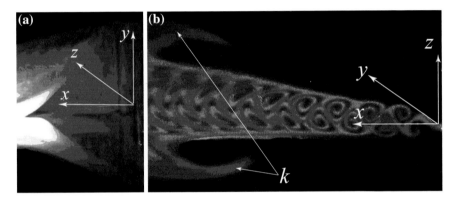

Fig. 8.22 Visualization of the plane microjet in transverse acoustic field ($f = 100$ Hz): general view of the jet in x-y plane (**a**) and its streamwise section at $y = 0$ mm (**b**); the microjet folding is marked by k (see Presentation Chap. 8: "Multimedia files Nos. 8.14–8.16") (http://extras.springer. com)

Fig. 8.23 Visualization of the plane microjet forced by transverse acoustic excitation ($f = 100$ Hz) (see Presentation Chap. 8: "Multimedia file No. 8.17") (http://extras. springer.com)

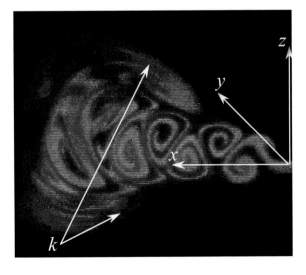

Key points

- Kelvin–Helmholtz instability still applies to round macrojets with a top-hat mean velocity profile at the nozzle exit at reduction of their diameter to about 5 mm.
- Evolution of microjets both with a top-hat and a parabolic mean velocity profile at the nozzle exit becomes completely different.
- Transformation of the round microjet into plane one caused by transverse acoustic forcing is revealed.
- Evolution of the round microjet is dominated by the mechanism of plane-jet sinusoidal instability.
- The phenomenon of microjet splitting into two secondary jets developing independently of each other is revealed.
- The secondary microjets are subjected to high-frequency instability.
- Sinusoidal instability is inherent in plane macro- and microjets forced by acoustic oscillations and in quiet environment.
- Pseudo plane ($l/h = 10$) and round microjets evolve similarly under transverse acoustic excitation.
- New phenomenon of the microjet folding at its edges in the direction of velocity vector created by transverse acoustic waves is demonstrated.
- Folding of the plane microjet results in its extinction and further transition to turbulence.
- Contrary to the instability of plane microjet, the sinusoidal instability of round microjet is affected by the direction of acoustic forcing.

References

Abramov OV, Borisov YY, Oganyan RA (1987) Critical sound pressure in the acoustic atomization of liquids. Sov. Phys. Acoust. 33:339–345

Boiko AV, Grek GR, Dovgal AV, Kozlov VV (2006) Physical mechanisms of the turbulent transition in the open flows. NIC "Regular and Chaotic Dynamics"; Institute of Computer Studies, Moscow, Izhevsk, pp 1–304 (in Russian)

Carpenter JB, Baillot F, Blaisot JB, Dumouchel C (2009) Behavior of cylindrical liquid jets evolving in a transverse acoustic field. Phys Fluids 21:023601–0236015

Gau C, Shen CH, Wang ZB (2009) Peculiar phenomenon of micro-free-jet flow. Phys Fluids 21:092001–092003

Grek GR, Kozlov VV, Kozlov GV, Litvinenko YuA (2009) Instability modeling of a laminar round jet with parabolic mean velocity profile at the nozzle exit. Byull. Novosibirsk State Univ. Ser. Fiz. 4(1):14–324 in Russian

Heister SD, Rutz MW, Hilbing JH (1997) Effect of acoustic perturbation on liquid jet atomization. J Propul Power 13:82–97

Hoover DV, Ryan HM, Pal S, Merkle CL, Jacobs HR, Santoro RJ (1991) Pressure oscillation effects on jet breakup. ASME, Heat Mass Transfer Spray Syst HTD 187:27–41

Kozlov GV, Grek GR, Sorokin AM, Litvinenko YuA (2008a) Influence of the initial conditions at the nozzle exit on a flow structure and stability of the plane jet. Byull. Novosibirsk State Univ. Ser. Fiz. 3(3):25–37 in Russian

Kozlov VV, Grek GR, Sorokin AM, Litvinenko YuA (2008b) Influence of initial conditions at the nozzle exit on a structure and development characteristics of a round jet. Thermophys Aeromech 15(1):55–68

Kozlov VV, Grek GR, Kozlov GV, Litvinenko YuA (2009) Physical aspects of subsonic jet flows evolution. In: The collection of proceedings "Successes of mechanics of continuum" to the 70 anniversary of academician VA Levin, Dalnauka, Vladivostok, pp 331–351 (in Russian)

Lim TT, New TH, Luo SC (2001) On the development of large scale structures of a jet normal to a cross-flow. Phys Fluids 3:770–775

Litvinenko MV, Kozlov VV, Kozlov GV, Grek GR (2004) Effect of streamwise streaky structures on turbulization of a circular jet. J Appl Mech Tech Phys 45(3):349–357

Sato H (1960) The stability and transition of a two-dimensional jet. J Fluid Mech 7(1):53–80

Yu MH, Monkewitz PA (1993) Oscillations in the near field of a heated two-dimensional jet. J Fluid Mech 255:323–347

Chapter 9
Stability of Subsonic Macro- and Microjets and Combustion

In this chapter, results of experimental studies on stability and combustion of round and plane propane macro- and microjets at low Reynolds numbers in transverse acoustic field are considered. Some features of flame evolution are shown. New phenomena in combustion of round and plane propane microjets forced by transverse acoustic waves are discovered and explained. Additional material to this Chapter is given in multimedia files supplementing the book entitled as Chap. 9: "Multimedia files Nos. 9.1–9.10" (http://extras.springer.com).

9.1 Diffusion Combustion of a Gaseous Fuel (Propane) in a Round Macrojet

9.1.1 Laminar and Turbulent Round Macrojets with a Top-Hat Mean Velocity Profile at the Nozzle Exit

The experimental setup for generation and investigation of a round macrojet 5 mm in diameter is schematically shown in Fig. 9.1. The macrojet has a top-hat mean velocity profile at the nozzle exit (Fig. 9.2).

As in the study of a round macrojet 20 mm in diameter (Litvinenko et al. 2004), the technique of flow visualization through synchronization of the Kelvin–Helmholtz vortices with actuation of laser illumination of the flow by acoustic oscillations is used. Streaky structures are generated by means of roughness elements (5 pieces) glued onto the inner surface of the nozzle. The visualization data (Fig. 9.3) clearly show the Kelvin–Helmholtz vortices and the result of their interaction with streaky structures, leading to star-shaped distortion of ring vortices in the form of five azimuthal beams corresponding to positions of streaky structures.

© The Author(s) 2016
V.V. Kozlov et al., *Visualization of Conventional and Combusting
Subsonic Jet Instabilities*, SpringerBriefs in Applied Sciences and Technology,
DOI 10.1007/978-3-319-26958-0_9

Fig. 9.1 Experimental
arrangement: setup for
generation of a round
macrojet with a top-hat mean
velocity profile at the nozzle
exit

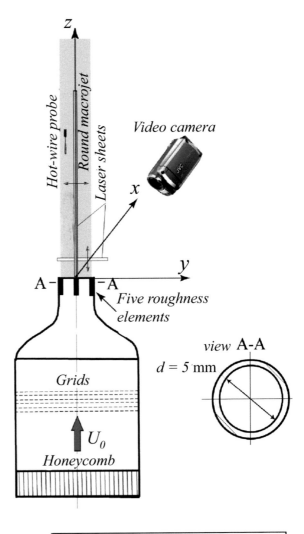

Fig. 9.2 Mean velocity
profiles in round macrojet
cross sections at different
distances from the nozzle exit
(*1, 2, 3*—5, 10, and 15 mm,
respectively)

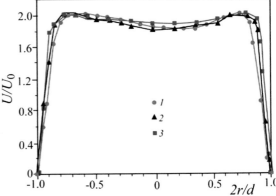

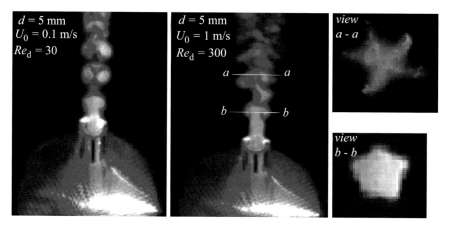

Fig. 9.3 Visualization of a laminar round macrojet with a top-hat mean velocity profile at the nozzle exit for different flow velocities and jet cross sections at interaction of the jet with streaky structures generated by roughness elements

Experimental studies of propane macrojet combustion show that laminar propane jet combustion with a small jet velocity leads to the formation of the flame attached to the nozzle exit. As the jet velocity increases, it is possible to observe turbulent propane jet flow combustion with the flame lifted above the nozzle exit (Fig. 9.4). Deformation of the first ring vortex caused by the existence of streaky structures can be seen in this pattern. Apparently, the flame propagates downstream beginning from the first coherent structure, which indicates that combustion does

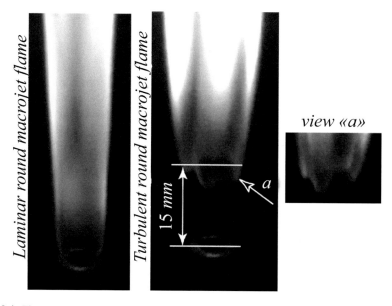

Fig. 9.4 Flame patterns of laminar and turbulent combustion of a round propane macrojet (see Presentation Chap. 9: "Multimedia file No. 9.1") (http://extras.springer.com)

not affect the structural characteristics of the round jet that are observed without combustion (ring vortices, streaky structures, and result of their interaction).

9.1.2 Laminar and Turbulent Round Macrojets with a Parabolic Mean Velocity Profile at the Nozzle Exit

The experimental setup for generation and investigation of a round macrojet 5 mm in diameter is schematically shown in Fig. 9.5. The macrojet has a parabolic mean velocity profile at the nozzle exit (Fig. 9.6).

Preliminary experimental studies of propane macrojet combustion showed that laminar propane jet combustion with a small jet velocity leads to formation of the flame attached to the nozzle exit. As the jet velocity increases, it is possible to observe turbulent propane jet flow combustion with the flame lifted above the nozzle exit (Fig. 9.7). A ring structure is seen in this turbulent flame pattern, which demonstrates that the flame propagates in a narrow region at the jet periphery.

9.2 Diffusion Combustion of a Gaseous Fuel (Propane) in a Round Microjet with Acoustic Forcing

9.2.1 Round Microjet in a Transverse Acoustic Field (Flame Flattening)

Experimental studies of the structure and characteristics of the development of a round macrojet (at the nozzle exit diameter of 1.5 mm) and microjet (at the nozzle exit diameter from 200 to 500 μm) under the influence of a high-intensity (90–100 dB) transverse acoustic field show that the round microjet becomes flattened and starts to perform sinusoidal oscillations as a whole, followed by a bifurcation further downstream (Kozlov et al. 2010, 2011; Litvinenko et al. 2011). Experimental studies of diffusion propane round microjet combustion were conducted at acoustic forcing. The experimental arrangement is schematically shown in Fig. 9.8.

Hot-wire measurements of the mean (U) and fluctuating (u') streamwise velocity distributions in the cross section of the round jet at a distance of 0.3 mm from the nozzle exit with the jet velocity at the flow axis $U_0 = 4$ m/s reveal the existence of a top-hat mean velocity profile if the nozzle channel is sufficiently short (Fig. 9.9) and a parabolic mean velocity profile (Fig. 9.10) if the aspect ratio of the nozzle is large ($l/d \approx 40$, where l is the nozzle channel length and d is the nozzle exit diameter).

It should be noted that we managed to measure and find the main difference (between the top-hat and parabolic mean velocity profiles) in the distribution of the streamwise velocity components in the cross section of a round jet with such a small

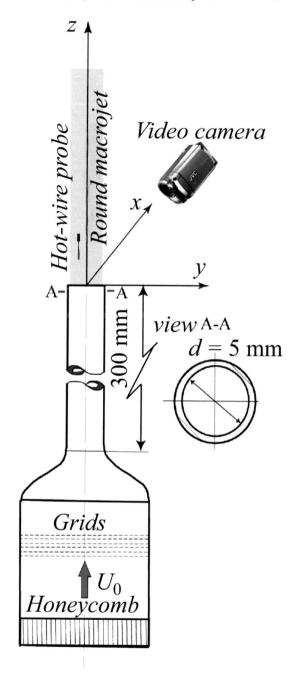

Fig. 9.5 Experimental arrangement: setup for generation of a round macrojet with a parabolic mean velocity profile at the nozzle exit

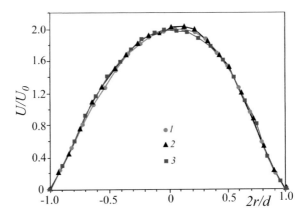

Fig. 9.6 Parabolic mean velocity profiles for round macrojet cross sections at different distances from the nozzle exit (*1, 2, 3*—5, 10, and 15 mm, respectively)

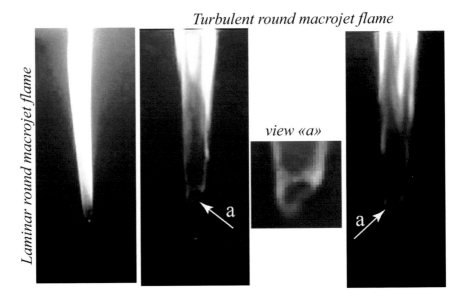

Fig. 9.7 Flame patterns of laminar and turbulent round propane macrojet combustion (see Presentation Chap. 9: "Multimedia file No. 9.2") (http://extras.springer.com)

diameter ($d = 1$ mm) for the first time. A hot-wire probe with a sensitive element 0.3 mm long was used. The step of measurements was 0.02 mm, and time averaging of the sensor readings at each measurement point was applied.

The plots in Fig. 9.9 reveal a peak of the velocity fluctuation intensity ($u' \approx 0.4$ % of U_0) in regions of the maximum velocity gradient and its minimum ($u' \approx 0.2$ % of U_0) in jet core regions for the jet with a top-hat mean velocity profile at the nozzle exit. On the other hand, the plots in Fig. 9.10 demonstrate a peak of

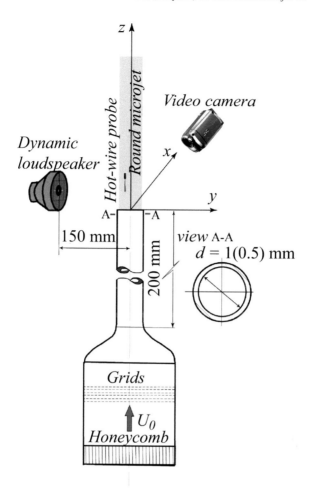

Fig. 9.8 Experimental arrangement

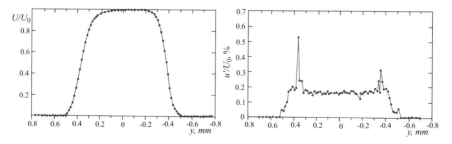

Fig. 9.9 Mean (U) and fluctuating (u') velocity distributions in the round microjet cross section at a distance of 3 mm from the nozzle exit for a short nozzle, $U_0 = 4$ m/s

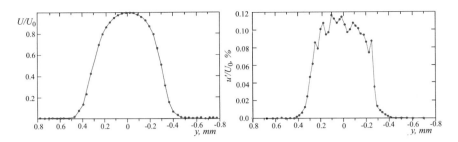

Fig. 9.10 Mean (*U*) and fluctuating (*u′*) velocity distributions in the round microjet cross section at a distance of 3 mm from the nozzle exit for a long nozzle (*l/d* ≈ 40), U_0 = 4 m/s

the velocity fluctuation intensity ($u′$ ≈ 0.1 % of U_0) at the jet axis for the jet with a parabolic mean velocity profile at the nozzle exit. These results are consistent with the data of measurements of the mean and fluctuating velocity profiles for round macrojets (Litvinenko et al. 2004). However, in contrast to the round macrojet with a top-hat mean velocity profile at the nozzle exit, where the development of the round jet is associated with the Kelvin–Helmholtz instability accompanied by generation of ring vortices with varicose instability of streaky structures in near-wall shear flows, the parabolic macrojet and microjet both with top-hat and parabolic mean velocity profiles at the nozzle exit are subjected only to sinusoidal instability of the entire jet. Attempts to excite varicose mode instability (ring vortices) in these flows failed. Experiments show that the macrojet with a parabolic mean velocity profile at the nozzle exit retains a purely laminar flow of a large range without ring vortices. The mechanism of turbulent breakdown of the jet is connected with other instabilities. On the other hand, the round microjet both with top-hat and parabolic mean velocity profiles at the nozzle exit also retains a purely laminar flow of a large range without ring vortices.

On the basis of the gained knowledge about the characteristics of the evolution of round macro- and microjets both with and without acoustic forcing, diffusion combustion of a propane round microjet with a parabolic mean velocity profile at the nozzle exit was experimentally studied. The flame evolution was registered by a digital video camera synchronized with the frequency of the acoustic effect on the jet. The frequency of acoustic forcing was varied from several Hz units to 4 kHz, and the sound pressure level was 90 dB. The jet velocity (U_0 = 16.6 m/s) was controlled by a precision flowmeter.

The lifted flames in the case of propane round jet combustion both with and without the transverse acoustic field are shown in Fig. 9.11. It is seen that a lifted flame is considerably extended (almost by four times) under the influence of a transverse acoustic field as compared to a flame without acoustic forcing. The question is what is the reason for this difference. In the works of (Kozlov et al. 2010, 2011), it is shown that the round microjet under the influence of a transverse acoustic field is flattened and acquires features typical for the plane jets development, for example, sinusoidal oscillations of the entire jet (Fig. 9.12).

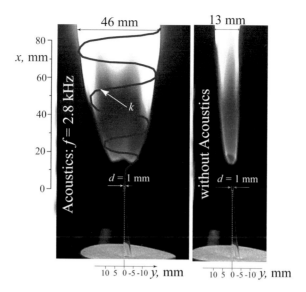

Fig. 9.11 Flame propagation along a flattened and sinusoidally oscillating round microjet in a transverse acoustic field during propane round jet diffusion combustion (k denotes sinusoidal oscillations of the flattened round microjet under acoustics forcing, $f = 2.8$ kHz, and $U_0 = 16.6$ m/s) (see Presentation Chap. 9: "Multimedia file No. 9.3") (http://extras.springer.com)

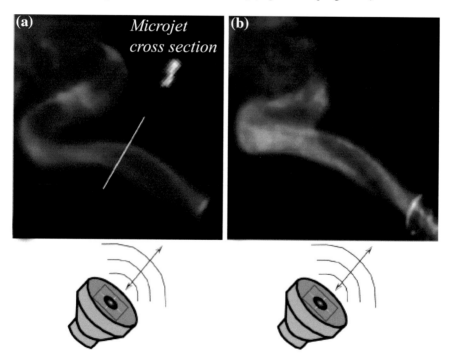

Fig. 9.12 Visualization of the round microjet evolution and flattening caused by a transverse acoustic field: $f = 40$ Hz (**a**) and 100 Hz (**b**), the sound pressure level is 90 dB (the figure is taken from the paper by Kozlov et al. 2011)

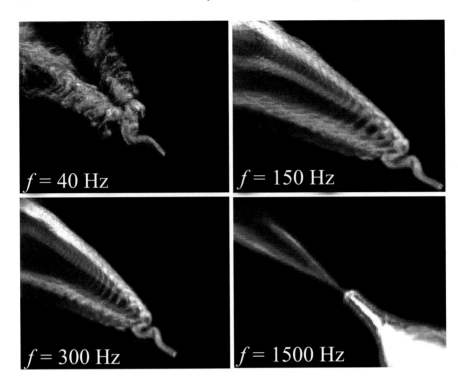

Fig. 9.13 Visualization of the round microjet bifurcation at different frequencies of the transverse acoustic field, the sound pressure level is 90 dB (the figure is taken from the paper by Kozlov et al. 2011)

It should be noted that sinusoidal oscillations of the entire plane jet (Fig. 9.12) results in natural broadening of the jet in the plane perpendicular to the direction of the transverse acoustic field and, finally, in the jet bifurcation into two secondary jets (Fig. 9.13). In the situation of propane round microjet combustion shown in Fig. 9.11, the dynamics of the flame evolution resembles the process of round jet flattening in a transverse acoustic field. The leading edge of the lifted flame without acoustic excitation clearly demonstrates the occurrence of a distorted ring-type structure, which means that the flame propagates only along the jet surface. On the other hand, the jet flame under the influence of a transverse acoustic field (Fig. 9.11) shows the absence of ring-type structures in the lifted flame configuration and powerful flame broadening (approximately by three times) further downstream due to sinusoidal oscillations of the entire jet.

9.2.2 Round Microjet in a Transverse Acoustic Field (Bifurcation of the Jet)

Considering the propane jet flame behavior further downstream (Fig. 9.14, nozzle diameter $d = 1$ mm), we can clearly see the flame bifurcation, which is directly connected with the mechanism observable in the case of the round jet splitting under the transverse acoustic excitation (Fig. 9.13).

A similar result of the flame bifurcation during diffusion combustion of a propane round microjet with the nozzle diameter $d = 0.5$ mm is shown in Fig. 9.15.

The flame bifurcation is observed, but at the acoustic forcing frequency approximately twice larger than in the case shown in Fig. 9.14. Thus, the acoustics forcing frequency leading to the flame bifurcation during propane round microjet combustion increases as the nozzle exit diameter decreases. It should be noted that, in the absence of acoustic forcing, the flame propagating on the round microjet surface during diffusion combustion of propane retains the configuration typical of the jet with a parabolic mean velocity profile at the nozzle exit. This configuration is characterized by the absence of coherent vortex structures, such as ring vortices, and by the occurrence of only a laminar jet flow of a large range (Chap. 4, Kozlov et al. 2008). This flame configuration remains unchanged in situations with both attached and lifted flames.

In our case, the flame bifurcation in the transverse acoustic field was observed only for the lifted flame. On the contrary, the attached flame bifurcation of round

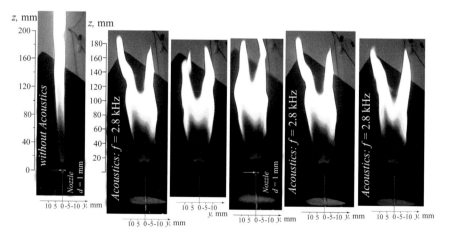

Fig. 9.14 Flame bifurcation under the transverse acoustic forcing in the case of propane round microjet ($d = 1$ mm) diffusion combustion (on the *left*—flame without acoustics, on the *right*—photos of the flame bifurcation due to the transverse acoustic field, $f = 2.8$ kHz, $U_0 = 16.6$ m/s) (see Presentation Chap. 9: "Multimedia file No. 9.4") (http://extras.springer.com)

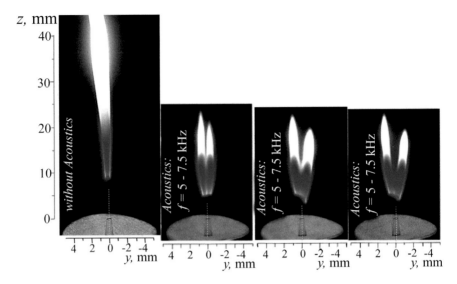

Fig. 9.15 Flame bifurcation under the transverse acoustic forcing in the case of propane round microjet ($d = 0.5$ mm) diffusion combustion (on the *left*—flame without acoustics, on the *right*—photos of the flame bifurcation due to the transverse acoustic field, $f = 5$–7.5 kHz, $U_0 = 12.5$ m/s) (see Presentation Chap. 9: "Multimedia files Nos. 9.5, 9.6") (http://extras.springer.com)

and plane jets under acoustic forcing was observed in (Suzuki et al. 2011). On the other hand, the authors of the work (Krivokorytov et al. 2012) also found the flame bifurcation in a lifted round microjet under acoustic excitation. The following circumstance should also be noted. The mechanism of the round jet flame bifurcation under the influence of acoustic oscillations is not clear from the previous work (Suzuki et al. 2011; Krivokorytov et al. 2012). We believe that the answer to this question can be given on the basis of results of our studies of the influence of a transverse acoustic field on the round microjet behavior in the absence of combustion (Kozlov et al. 2010). It is round microjet flattening under the acoustic forcing that transforms it to a plane quasi-two-dimensional jet subjected to sinusoidal oscillations, which result in jet breakdown into two parts further downstream (Kozlov et al. 2010). It is also found that sinusoidal instability of the round microjet flame, i.e. its flattening and sinusoidal oscillations, depends on the direction of the flow velocity vector created by the transverse acoustic field.

9.3 Diffusion (Propane) and Premixed (Propane/Air) Combustion in a Plane Microjet in a Transverse Acoustic Field

9.3.1 Diffusion Combustion of a Propane Plane Microjet in a Transverse Acoustic Field (Jet and Flame Bifurcation)

Experimental studies of the structure and characteristics of the development of a plane macrojet (the nozzle width is $h = 2.5$ mm) and microjet (the nozzle width is $h = 700$, 500, and 200 μm) under the influence of a high-intensity (90–100 dB) transverse acoustic field were carried out. These investigations show that the jets are subjected to sinusoidal oscillations and to a bifurcation into two jets further downstream (Chap. 6; Kozlov et al. 2010, 2011; Litvinenko et al. 2011). Diffusion combustion of a propane plane microjet with a small aspect ratio ($l/h = 10$, where l is the nozzle length and h is the nozzle width) under the transverse acoustic forcing was experimentally studied (Grek et al. 2013). The experimental setup is schematically shown in Fig. 9.16.

On the basis of the obtained data on the characteristics of the development of plane macro- and microjets both in the absence and presence of the transverse acoustic field, diffusion combustion and combustion of a premixed propane/air mixture of the plane microjet with a parabolic mean velocity profile at the nozzle exit was experimentally studied. The parabolic mean velocity profile at the nozzle exit was created with a sufficiently large aspect ratio of the nozzle channel ($l/h \geq 150$, where l is the length of the nozzle channel and h is the nozzle width) (Chap. 6; Kozlov et al. 2012). The downstream evolution of the flame was recorded by a digital video camera synchronized with the frequency of acoustic forcing. The frequency of acoustic forcing varied from several Hz units up to 6 kHz, and the sound pressure level was approximately 90 dB. The velocity on the jet axis in two experiments ($U_0 \approx 20.8$ m/s and 32 m/s) was controlled by a precision flowmeter.

The attached flame patterns both with and without the influence of transverse acoustic waves are shown in Fig. 9.17.

It is seen that the attached flame under the acoustic forcing is appreciably expanded in the xz-plane downstream due to broadening of the sinusoidal vortex street generated by acoustic oscillations. Under the excitation, the transverse size of the flame in the xz-plane continuously grows downstream (Kozlov et al. 2010, 2011; Litvinenko et al. 2011). As the jet velocity grows and reaches a certain threshold value, the flame sharply departs from the nozzle exit and the so-called lifted flame is realized (Fig. 9.18). In this situation, the flame bifurcation at the acoustic influence in the frequency range from 1 to 3 kHz is clearly observed (Fig. 9.18). The flame in the absence of acoustic forcing is shown in Fig. 9.18. It is seen that the flame lift-off height from the nozzle exit is approximately twice larger than the flame lift-off height in the situation with acoustic forcing.

Fig. 9.16 Experimental setup

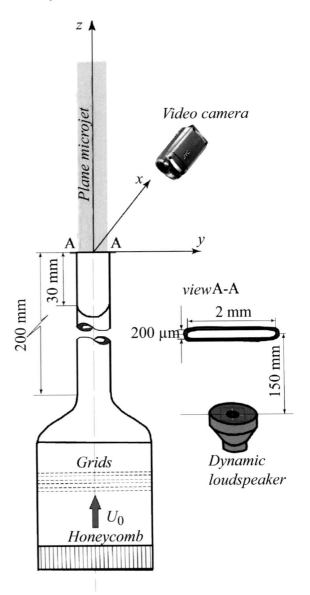

Apparently, acoustics stabilizes the process of combustion and promotes heat and mass transfer intensification between propane and air, which results in expansion of the combustion region, shifting of the flame to the nozzle exit, and, finally, its bifurcation into two flames. These two flames propagate over the surface of two jets arising as a result of acoustically induced breakdown of the sinusoidal vortex street of the plane microjet without combustion investigated earlier and presented in Chap. 8.

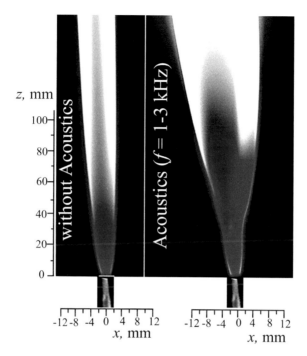

Fig. 9.17 Flame broadening under the transverse acoustic forcing in the case of propane plane microjet ($h = 0.2$ mm) diffusion combustion (on the *left*—natural flame, on the *right*—flame broadening due to the transverse acoustic excitation, $f = 1$–3 kHz, $U_0 \approx 20.8$ m/s) (see Presentation Chap. 9: "Multimedia file No. 9.7") (http://extras.springer.com)

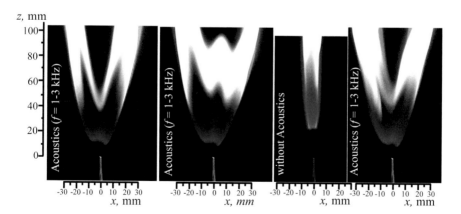

Fig. 9.18 Flame bifurcation in the transverse acoustic field in the case of propane plane microjet ($h = 0.2$ mm) diffusion combustion, $f = 1$–3 kHz, $A \approx 90$ dB, $U_0 \approx 32$ m/s) (see Presentation Chap. 9: "Multimedia file No. 9.7") (http://extras.springer.com)

9.3.2 Plane Microjet in a Transverse Acoustic Field at Propane Diffusion Combustion (Bifurcation of the Jet and Flame)

Experimental studies of propane diffusion combustion in a plane microjet at the nozzle exit with a large aspect ratio ($l/h = 180$, where l is the nozzle length and h is the nozzle width) were carried out with the acoustic influence on the flame (Grek et al. 2013). The experimental setup is shown in Fig. 9.19.

In this case, in contrast to the microjet emanating from the nozzle with a small aspect ratio ($l/h = 10$), it is possible to observe not only the process of the jet bifurcation under the transverse acoustic excitation, but also vortex sheet folding in the opposite directions in each period of acoustic forcing (Fig. 9.20b, c).

In the process of diffusion combustion of this jet, flame propagation on the surface of these two divided and twisting jets (Fig. 9.20a) can be seen. It is necessary to notice that the lifted flame has a deformed flame front due to nozzle exit roughness, which results in generation of small-scale coherent vortex structures. In the absence of acoustic forcing, the microjet flame in the yz-plane is steadier and it is not broadened in the downstream direction, which demonstrates the absence of oscillatory process (Fig. 9.20a).

It is interesting to consider the behavior of this plane microjet flame in the xz-plane, i.e. in the plane where the formation of a sinusoidal vortex street and its bifurcation under the transverse acoustic forcing actually occur. The photos of the microjet flame both with and without acoustic effect are shown in Fig. 9.21a. The smoke visualization pattern of this jet in the absence of combustion with the jet bifurcation under the influence of acoustic forcing is also demonstrated (Fig. 9.21b). It is possible to observe the flame bifurcation and its downstream broadening approximately twice as compared to the case without acoustic forcing (Fig. 9.21a).

Thus, the behavior of the flame of the plane microjet at the nozzle exit with a large aspect ratio ($l/h = 180$) during diffusion combustion of propane completely reflects the phenomena observable earlier in studying the evolution of this air microjet flow in the absence of combustion (jet bifurcation and its folding because of the tip effects).

9.3.3 Plane Microjet in a Transverse Acoustic Field at Combustion of a Premixed Propane/Air Mixture (Bifurcation of the Jet)

The flame evolution during premixed propane/air mixture combustion in a plane microjet under the transverse acoustic forcing was experimentally studied (Grek et al. 2013). The experimental setup is shown in Fig. 9.16. The propane/air ratio was 37.2 %/62.8 % (the equivalence ratio was $\varphi = 14$). At smaller equivalence

Fig. 9.19 Experimental
arrangement: setup for
generation of a plane microjet
($l = 36$ mm, $h = 0.2$ mm) with
a parabolic mean velocity
profile at the nozzle exit

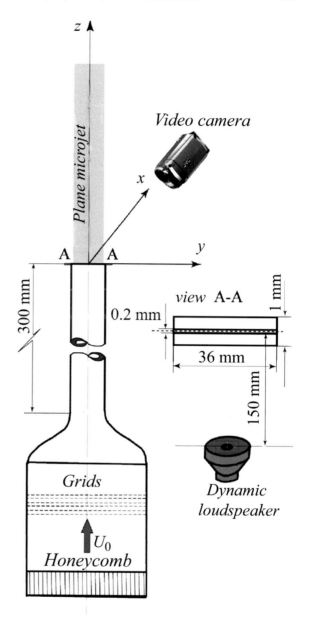

ratios, the transverse acoustic field produced a much weaker effect on jet combustion, if any. As in the situation of diffusion combustion of a plane propane
microjet (Fig. 9.18), the plane microjet flame in the case of combustion of a premixed propane/air mixture also experiences a bifurcation under acoustic forcing
(Fig. 9.22). The jet flame in the absence of the acoustic field and under the acoustic

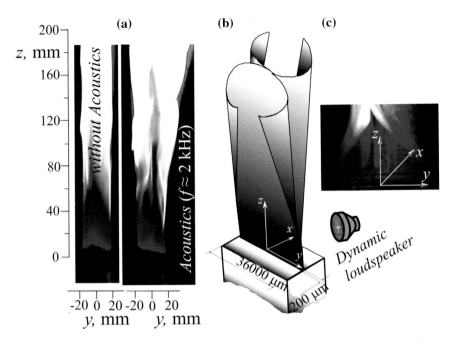

Fig. 9.20 Flame folding in the transverse acoustic field in the case of propane plane microjet ($l/h = 180$) diffusion combustion, $f = 2$ kHz, $A \approx 90$ dB (**a**), scheme of this process (**b**), and smoke visualization of the plane microjet without combustion (figure is taken from the paper by Kozlov et al. 2011) (**c**), $U_0 \approx 16$ m/s (see Presentation Chap. 9: "Multimedia file No. 9.8") (http://extras.springer.com)

effect at frequencies smaller than 500 Hz is attached to the nozzle. It is known that the lift-off of the flame from the nozzle exit occurs as the jet velocity increases or, as in this case, when the jet velocity remains constant; flame separation from the nozzle exit occurs owing to the acoustic influence at certain frequencies ($f = 1$–3 kHz). In the situation considered here (Fig. 9.22), the lifted flame is sharply broadening in the xz-plane, and its bifurcation can be observed further downstream, as well as in the previous case of diffusion combustion of the plane microjet subjected to acoustic forcing (Fig. 9.18). One more interesting phenomenon can be noted: a sharp increase in the flame lift-off height from the nozzle exit (approximately by three times) after termination of acoustic forcing (Fig. 9.22, the most right picture). It is also found that instability of the plane microjet flame is independent of the direction of the transverse acoustic field vector, in contrast to its dependence on the direction of the transverse acoustic field vector for the round microjet flames.

Thus, experimental studies on diffusion and premixed propane/air mixture combustion of subsonic plane microjets show that the flame development reflects all phenomena that were found earlier in studying the mechanism of the development of plane macro- and microjets in a transverse acoustic field, but in the absence

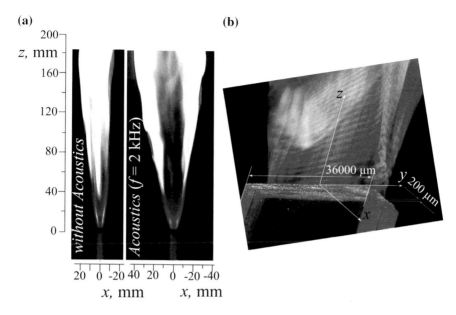

Fig. 9.21 Flame bifurcation in the transverse acoustic field in the case of propane plane microjet ($l/h = 180$) diffusion combustion, $f = 2$ kHz, $A \approx 90$ dB (**a**) and smoke visualization of the microjet without combustion (figure is taken from the paper by Kozlov et al. (2011)) (**b**), $U_0 \approx 16$ m/s (see Presentation Chap. 9: "Multimedia file No. 9.9") (http://extras.springer.com)

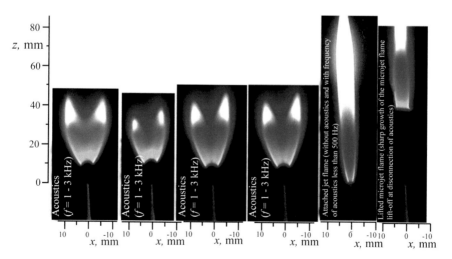

Fig. 9.22 Flame bifurcation under the transverse acoustic forcing in the case of premixed propane/air mixture combustion in a plane microjet ($l = 2$ mm, $h = 0.2$ mm), $f = 13$ kHz, $A \approx 90$ dB, $U_0 \approx 14$ m/s (see Presentation Chap. 9: "Multimedia file No. 9.10") (http://extras.springer.com)

of combustion. They include excitation of a sinusoidal vertical stress by sound, broadening and bifurcation of the vortex street further downstream (Chap. 8; Kozlov et al. 2010, 2011, 2012).

9.3.4 Analysis of One Possible Reason for the Mechanism of the Flame Bifurcation During Diffusion Combustion of a Propane Round Jet

The characteristics of an acoustically excited diffusion flame with a focus on studying the flame bifurcation mechanism were investigated in the work (Suzuki et al. 2011). The authors of this work tried to understand the bifurcation mechanism with the help of an experiment on diffusion combustion of a plane jet. Though the round jet bifurcation mechanism is not clear yet, the authors of the work (Suzuki et al. 2011) considered one possible explanation of this phenomenon: the effect of the baroclinic torque (Polifke 2012), which is proportional to the vector of the vector product of the pressure and density gradients (Fig. 9.23).

The pressure gradient is generated by a horizontally propagating sound wave, while the density gradient is caused by distributions of the gas composition and temperatures, which are basically axisymmetric around the vertical axis of a round nozzle. The baroclinic torque distributed in the flow field is, thus, three-dimensional. The authors of the work (Suzuki et al. 2011) replaced a round (*3D*) burner by a plane (*2D*) burner. The flames of the round and plane burners were compared [see Fig. 6 from (Suzuki et al. 2011)]. Experimental results show that the microjet bifurcation phenomenon is absolutely two-dimensional. It is demonstrated that the density gradients in the direction from the jet axis cannot be a key factor of the jet bifurcation if this phenomenon is caused by the baroclinic torque effect. Thus, the phenomenon of the round microjet bifurcation in a transverse acoustic field, apparently, is associated with some other mechanism, namely, with round microjet flattening in a transverse acoustic field, as it is shown in Fig. 9.12.

In this situation, the jet is subjected to sinusoidal oscillations as a whole (Figs. 9.12 and 9.13). The flame propagating over such a jet bifurcates downstream due to sinusoidal oscillations of the entire jet as a whole and its separation into two

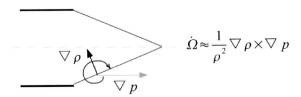

Fig. 9.23 Vorticity generated due to baroclinic torque. A combination of misaligned pressure and density gradients causes generation of vorticity at the flame front (Polifke 2012)

jets in each half-cycle of oscillations (Figs. 9.14 and 9.15). Sinusoidal oscillations of the plane jet flame can be clearly observed in Fig. 3b, and its bifurcation is clearly seen in Fig. 3a from Suzuki et al. (2011). A similar result of the bifurcation of the round (Figs. 9.14 and 9.15) and plane (Figs. 9.18, 9.20, 9.21 and 9.22) microjets under the influence of a transverse acoustic field is obtained in the present experiments. It should be noted that, in contrast to (Suzuki et al. 2011) where jet combustion with the flame attached to the nozzle exit is presented, the lifted flame evolution is investigated in the present experiments. This phenomenon was observed for the acoustic forcing of both round and plane microjet.

Thus, the results of experimental studies on the influence of a transverse acoustic field on physical processes accompanying propane combustion in round and plane microjets at low Reynolds numbers are presented in this chapter. Some new phenomena concerning the influence of the transverse acoustic field on the flame evolution during propane microjet combustion are revealed. All these phenomena were observed earlier in studying the microjet development without combustion. Unfortunately, understanding of only physics of microjet combustion cannot give an answer to the question whether acoustics affects the fuel combustion completeness and whether it involves reduction of harmful emissions into the atmosphere. To answer this question, it is necessary to study the chemical processes of combustion, which is planned to be done together with the Institute of Chemical Kinetics and Combustion of the Siberian Branch of the Russian Academy of Sciences.

Key points

- It has been found that diffusion combustion of a laminar propane round macrojet with a top-hat mean velocity profile at the nozzle exit is accompanied by an attached flame developing downstream without pronounced pulsations.
- It has been shown that diffusion combustion of a turbulent propane round macrojet is characterized by flame liftoff from the nozzle exit, deformation of the first ring-shaped vortex due to streaky structures, high-frequency oscillations, and flame broadening.
- It has been found that diffusion combustion of a laminar propane round macrojet with a parabolic mean velocity profile at the nozzle exit is accompanied by the presence of an attached flame developing downstream smoothly.
- It has been shown that diffusion combustion of a turbulent propane round macrojet is accompanied by flame lift-off and by the presence of a ring vortex indicating flame propagation in a narrow region at the jet periphery, jet broadening, and high-frequency oscillations of the jet.
- It has been found that the round microjet combustion flame is subjected to flattening and bifurcation under the transverse acoustic excitation.
- It has been shown that the lifted flame during diffusion combustion of a round propane microjet in a transverse acoustic field is subjected to flattening and transverse broadening (by more than three times) as compared to the flame without acoustic forcing.

- It has been revealed that the lifted flame in the case of diffusion combustion of a round propane microjet is subjected to bifurcation due to the breakdown of the sinusoidal oscillating jet into two parts in each half-cycle of its oscillation in a transverse acoustic field.
- It has been found that the influence of a transverse acoustic field on diffusion combustion of propane and premixed propane/air mixture combustion in a plane microjet at small aspect ratios of the nozzle ($l/d = 10$) results in broadening of the combustion region and the flame bifurcation further downstream.
- It has been shown that the influence of a transverse acoustic field on the process of diffusion propane plane microjet combustion at large aspect ratios of the nozzle ($l/d = 180$) results in broadening of the combustion region, flame bifurcation, and flame propagation on the surface of each of the two secondary jets.
- It has been found that the mechanism of the flame round microjet bifurcation is associated with jet flattening in a transverse acoustic field. As a result, the jet development is two-dimensional, and it is impossible to explain the jet bifurcation phenomenon within the framework of the baroclinic torque effect.

References

Grek GR, Katasonov MM, Kozlov VV, Korobeynichev OP, Litvinenko YuA, Shmakov AG (2013) Features of the round and plane macro- and microjets combustion in a transverse acoustic field at the small Reynolds numbers. Byull Novosibirsk State Univ Ser Fiz 8(3):98–119 (in Russian)

Kozlov VV, Grek GR, Sorokin AM, Litvinenko YuA (2008) Influence of initial conditions at the nozzle exit on a structure and development characteristics of a round jet. Thermophys Aeromechanics 15(1):55–68

Kozlov VV, Grek GR, Litvinenko YuA, Kozlov GV, Litvinenko MV (2010) Subsonic round and plane macro- and microjets in a transverse acoustic field. Byull Novosibirsk State Univ Ser Fiz 5(2):28–42 (in Russian)

Kozlov VV, Grek GR, Litvinenko YuA, Kozlov GV, Litvinenko MV (2011) Round and plane jets in a transverse acoustic field. J Eng Thermophys 20(3):1–18

Kozlov VV, Grek GR, Litvinenko YuA, Kozlov GV, Litvinenko MV (2012) Influence of initial conditions at the nozzle exit and acoustical action on the structure and stability of a plane jet. Vis Mech Process 2(3):1–29

Krivokorytov MS, Golub VV, Volodin VV (2012) Acoustic oscillations effect on the methane diffusion combustion. Lett Tech Phys J (TPJ) 38(10):57–63 (in Russian)

Litvinenko MV, Kozlov VV, Kozlov GV, Grek GR (2004) Effect of streamwise streaky structures on turbulization of a circular jet. J Appl Mech Tech Phys 45(3):349–357

Litvinenko YuA, Grek GR, Kozlov VV, Kozlov GV (2011) Subsonic round and plane macro and microJet in a transverse acoustic field. Dokl Phys 436(1):58–63

Polifke W (2012) Influence of oscillating pressure gradient on premixed flame dynamics, project EU FP-7 ITN MYPLANET. http://129.187.45.233/tum-td/de/forschung/bereiche/themen/stanko. Accessed 28 June 2015

Suzuki M, Ikura S, Masuda W (2011) Comparison between acoustically-excited diffusion flames of tube and slit burners. In: Proceedings of the 11th Asian symposium on visualization, Niigata, Japan, pp 1–6

Printed in the United States
By Bookmasters